POCKET ANATOMY *of the* MOVING BODY

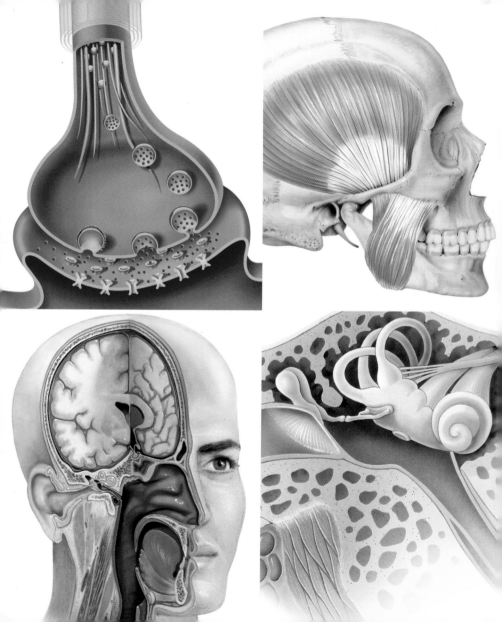

POCKET **ANATOMY** *of the*
MOVING BODY
The Compact Guide to the Science of Human Locomotion

CONSULTING EDITOR: **PROFESSOR JOHN BREWER**

First edition for North America published in 2017 by Barron's Educational Series, Inc.

A Global Book

© 2016 Quarto Publishing PLC, 6 Blundell Street, London N7 9BH, UK

Conceived, designed, and produced by Global Book Publishing

Illustrations (anatomical): Joanna Culley BA(Hons) RMIP, MMAA, IMI (Medical-Artist.com)

Illustrations (graphic): Robert Brandt

Additional illustrations: David Carroll, Peter Child, Deborah Clarke, Geoff Cook, Marcus Cremonese, Beth Croce, Hans De Haas, Wendy de Paauw, Levant Efe, Mike Golding, Mike Gorman, Jeff Lang, Alex Lavroff, Ulrich Lehmann, Ruth Lindsay, Richard McKenna, Annabel Milne, Tony Pyrzakowski, Oliver Rennert, Caroline Rodrigues, Otto Schmidinger, Bob Seal, Vicky Short, Graeme Tavendale, Jonathan Tidball, Paul Tresnan, Valentin Varetsa, Glen Vause, Spike Wademan, Trevor Weekes, Paul Williams, and David Wood

All inquiries should be addressed to:
Barron's Educational Series, Inc.
250 Wireless Boulevard
Hauppauge, NY 11788
www.barronseduc.com

ISBN: 978-1-4380-0906-3

Library of Congress Control Number: 2016942768

Printed in China

9 8 7 6 5 4 3 2 1

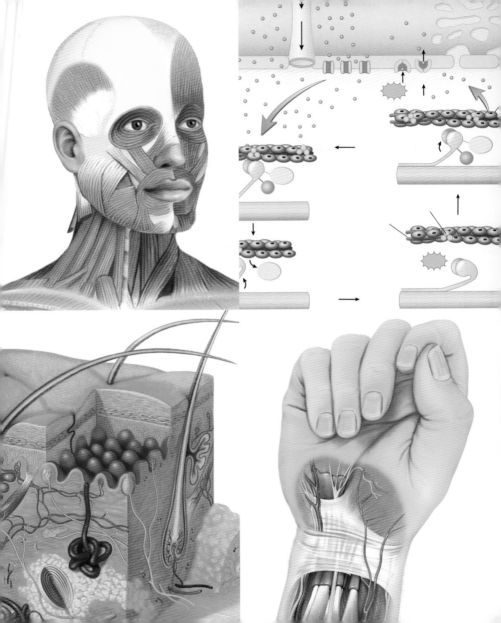

Contents

Introduction

THE HUMAN BODY IS DESIGNED TO MOVE. Without movement, people cannot undertake their daily lives. Furthermore, our species would not have survived, because without the ability to move our ancestors would not have been able to hunt for food and would have been easy prey for other species. While modern lifestyles, and the advances in technology that make them possible, have led to far less active lives for many, movement remains an essential part of our existence.

Humans are characterized by their vertical posture and bipedal locomotion—in other words, we stand and use two legs to move around. But of course, this is by no means the only manner in which we move, and it represents just a fraction of the various forms of movement that the human body is capable of. Many movements are subconscious; for example, breathing, blinking, the maintenance of body posture, and even the contraction of heart muscles, all

happen with little or no conscious effort. Furthermore, exercise and technological innovation have resulted in different opportunities for conscious movement that involve timing, skill, and the use of many different joints and muscles. Cycling and rowing are excellent examples of human movement that combine with technology to create means of human locomotion that in the past would not have been possible. Meanwhile, actions such as a golf swing, tennis serve, and kicking a ball all involve specific movements that have evolved as a result of sport and exercise becoming an integral part of the lives of many.

The human body is a highly complex organism. All its movements depend on interaction between the nervous system, bones, muscles, tendons, and ligaments, as well as on the supply of energy from respiration and fuel stores. Whereas basic and essential forms of movement are inherent and evolve and develop from birth, others can be improved through exercise and nutrition, or can decline through neglect. As the human body grows, many movements develop and become more refined. However, the inevitable decline in physiological function that accompanies aging makes many movements harder, and some are lost altogether. The fragility and complexity of movement is highlighted by the crucial role that the nervous system plays, transporting electrical impulses to the muscles to trigger contractions. Any damage to nerves that has a negative impact on the transmission of electrical signals can have a disastrous and potentially life-changing effect on our capacity for movement.

This book provides a comprehensive but easy-to-understand guide to the basic principles and science of human movement, including detailed descriptions of the main anatomical structures of the body, their function, and practical examples of human movement, all brought to life with high-quality images.

SECTION ONE:
THE HUMAN MUSCULOSKELETAL SYSTEM

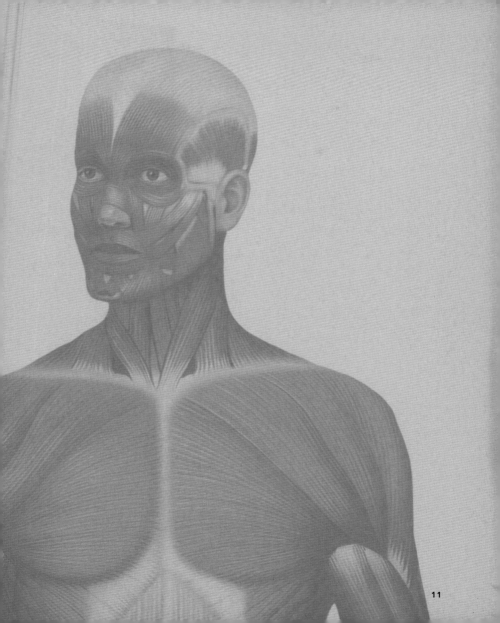

11

Chapter 1:
Introduction to the human frame

The human frame is designed to perform a number of different functions, often simultaneously and interdependently. In this section, we look at how the skeleton provides support and protection for fragile tissues and organs. This chapter also examines how the skeleton provides the scaffolding necessary for muscle attachment and acts as a lever to generate movement and maintain posture.

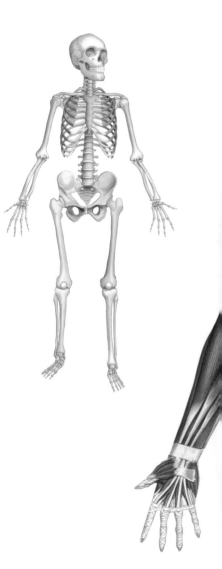

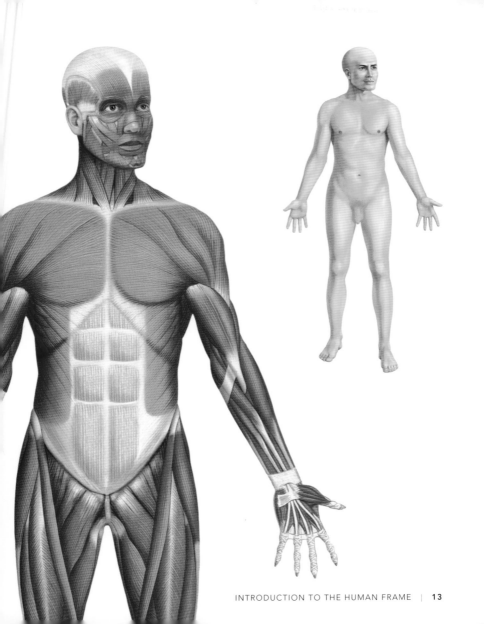

Skeletal system

Bones of the upper body—front

THE HUMAN SKELETON IS COMPOSED OF 270 BONES AT BIRTH—some of which fuse together as it matures and grows —and 206 bones in adulthood. The bones of the human skeleton are divided into two main categories: the axial skeleton (the bones of the head, ribs, and spinal column) and the appendicular skeleton (the bones of the rest of the human skeleton). The axial skeleton runs along the body's midline axis. It forms the upright axis of the body and is composed of the skull, the vertebral column, and the rib cage. The axial skeleton has a protective role, housing the brain, the spinal cord, and the organs within the chest cavity.

The skull is composed of 21 cranial and facial bones that are separate in childhood, allowing the skull and brain to grow. As the human body matures, these bones fuse together as one, thus providing more robust protection as an adult. The mandible, or jaw bone, connects with the temporal bone to form the only movable joint within the skull.

The thoracic cage is a bony and muscular structure that protects and supports the heart, lungs, and other structures. The sternum is a thin bone located along the midline of the anterior (front) side of the thoracic region of the skeleton. The sternum connects to the ribs via bands of cartilage called the costal cartilage. The ribs are long, flat bones that, together with the sternum, form the rib cage or thoracic cage. The sternum provides the only point of articulation for the pectoral girdle and upper limbs via the clavicle.

▶ **RIBS**

There are 12 pairs of ribs. The first seven pairs, known as "true ribs," attach to the sternum via costal cartilage. The lower five pairs are known as "false ribs," with three sharing a common cartilage connection to the sternum. The last two are termed "floating ribs," as they don't attach to the sternum.

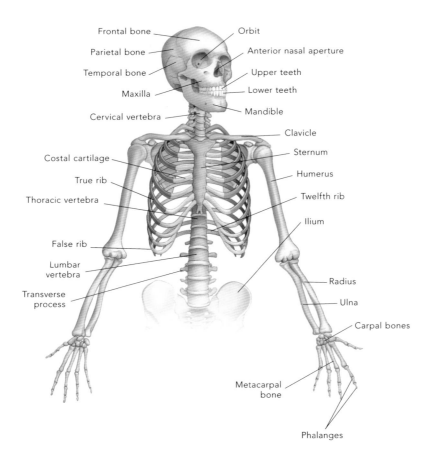

Frontal bone

Parietal bone

Temporal bone

Maxilla

Cervical vertebra

Costal cartilage

True rib

Thoracic vertebra

False rib

Lumbar vertebra

Transverse process

Orbit

Anterior nasal aperture

Upper teeth

Lower teeth

Mandible

Clavicle

Sternum

Humerus

Twelfth rib

Ilium

Radius

Ulna

Carpal bones

Metacarpal bone

Phalanges

Bones of the lower body—front

THE HUMAN LOWER LIMBS ARE SPECIFICALLY ADAPTED FOR BEARING WEIGHT IN A STANDING POSITION, maintaining an upright posture, and providing movement through walking and running.

The left and right coxae, or hip bones, comprise the pelvic girdle. They are formed through the fusion of three bones: the ilium, ischium, and pubis. They fuse in the region of the acetabulum, where the femur articulates to form the hip joint. The pelvis provides mobility during gait by rotating and allowing the legs to swing forward as well as protecting those organs housed in the pelvic girdle.

The femur is the longest and strongest bone in the body. The prominent rounded head articulates with the pelvis and is connected to the long shaft of the femur via the neck. Two tuberosities on the proximal shaft provide attachment sites for muscles that connect the thigh with the pelvis.

Anatomically, the leg is the part of the lower limb between the knee and the ankle. Two bones form this part of the limb: the tibia and the fibula. The tibia is by far the larger bone and articulates with the femur, while the fibula is much more delicate and mostly provides a site for the attachment of muscles.

The foot consists of seven tarsal bones, five metatarsal bones, and the phalanges of the toes. The largest of the tarsal bones is the calcaneus, which forms the heel of the foot and connects the talus and cuboid bones.

▶ **SEX DIFFERENCES**

The human skeleton exhibits subtle differences between males (shown here) and females, and the pelvis is one area where these differences can be observed. The female pelvis is larger and wider than the male pelvis and has a rounder pelvic inlet.

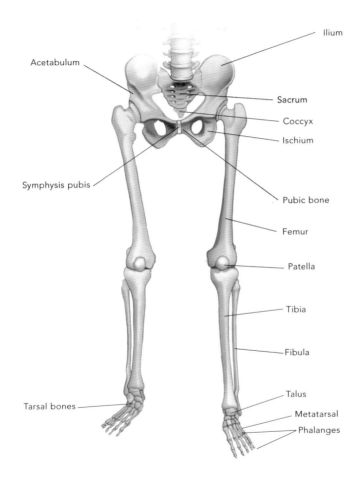

Ilium

Acetabulum

Sacrum

Coccyx

Ischium

Symphysis pubis

Pubic bone

Femur

Patella

Tibia

Fibula

Talus

Tarsal bones

Metatarsal

Phalanges

Skeletal system—rear view

THE VERTEBRAL COLUMN COMPRISES SEVEN CERVICAL VERTEBRAE IN THE NECK, twelve thoracic vertebrae in the chest region, five lumbar vertebrae in the lower back, as well as the fused sacral and coccygeal vertebrae. Each vertebra consists of a body, an arch, and a number of processes.

The vertebral column has multiple functions, including protecting the spinal cord, providing a site for muscle attachment, allowing a range of movement of the head and trunk, and supporting the weight of the head and trunk.

The scapula, or shoulder blade, is a flat, triangular bone that lies across the upper aspect of the thoracic cage. The scapula's only bony connection to the axial skeleton is with the clavicle and these together form the pectoral girdle. The scapula is suspended in place by a number of muscles, which allows for freedom of movement, but at the cost of some stability.

The upper limb is anatomically divided into the arm, between the shoulder and elbow, and the forearm, between the elbow and wrist. The arm consists of only the humerus, which articulates with the scapula to form the shoulder joint. Below the head of the humerus are two tubercles that provide attachment sites for muscles that move and stabilize the shoulder. Distally, the humerus articulates with both the radius and ulna to form the elbow joint.

The forearm consists of the radius and ulna, which articulate with eight carpal bones to form the wrist. Five metacarpals form the hand, with the digits made up by the phalanges.

▶ **ATLAS AND AXIS**

The first cervical vertebra, called the atlas, is a ring that does not have a vertebral body. It is attached to the second vertebral body, the axis, which acts as a pivot that the first vertebra rotates around. Most of the rotation in the neck is located in these top two segments.

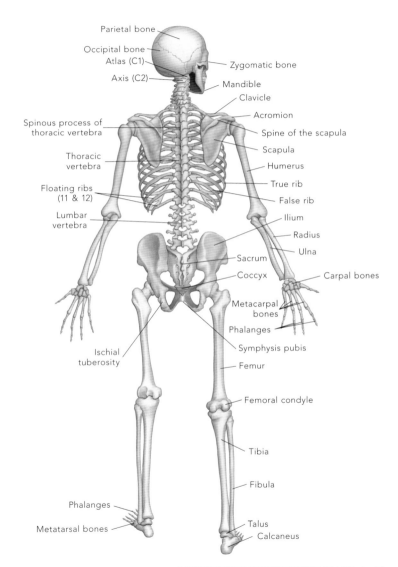

Parietal bone

Occipital bone

Atlas (C1)

Axis (C2)

Zygomatic bone

Mandible

Clavicle

Acromion

Spine of the scapula

Spinous process of
thoracic vertebra

Scapula

Thoracic
vertebra

Humerus

True rib

Floating ribs
(11 & 12)

False rib

Lumbar
vertebra

Ilium

Radius

Ulna

Sacrum

Coccyx

Carpal bones

Metacarpal
bones

Phalanges

Symphysis pubis

Ischial
tuberosity

Femur

Femoral condyle

Tibia

Fibula

Phalanges

Metatarsal bones

Talus

Calcaneus

Skeletal system—side view

AS A CONSEQUENCE OF UPRIGHT
HUMAN BIPEDAL POSTURE, THE CENTER
OF GRAVITY LIES SLIGHTLY BEHIND THE
HIP, which reduces the tendency for
gravity to pull the upper body forward.
During the initiation of gait, the center
of gravity is shifted forward, over the
feet, creating momentum.

Viewing the vertebral column from
the side, there are four apparent
curvatures. In the thoracic and sacral
regions, the spine curves posteriorly,
and these curves are present from early
embryonic development. In the
cervical and lumbar regions, the spine
is curved anteriorly. In the cervical
region of the spine, this curve becomes
apparent in early development when
the infant begins to raise his or her
head, whereas in the lumbar region,
the spinal curve develops when the
infant begins to sit upright and learns
to walk.

The pelvic girdle firmly attaches the
lower limbs to the axial skeleton via the
vertebral column. The sacroiliac joint
provides stability and weight transfer
from the axial skeleton to the lower
limbs. Supported by ligaments, the
sacroiliac joint transfers most of the
weight of the upper body to the legs.
This load is then transferred to the
large femur bones, then through the
lower limbs to the feet. Though there
are two bones in the lower leg, the tibia
performs the majority of the weight
bearing. The talus bone articulates with
the tibia and fibula to form the ankle
joint, transferring the weight of the
body to the foot.

▶ **CURVATURE**

Excessive or abnormal curvature of
the spine reflects spinal disease.
Exaggerated curvature in the thoracic
spine is called kyphosis. Exaggerated
lordotic curvature of the lumbar region
is known as lordosis. Scoliosis, lateral
curvature, is the most common
abnormal curvature.

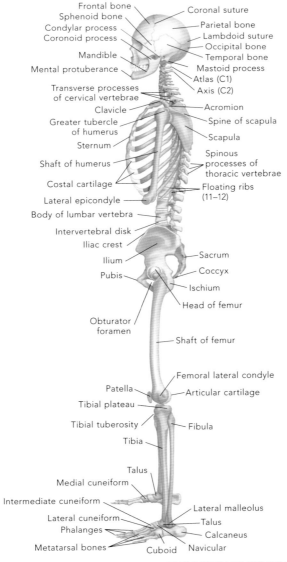

Frontal bone
Sphenoid bone
Condylar process
Coronoid process
Mandible
Mental protuberance
Transverse processes
of cervical vertebrae
Clavicle
Greater tubercle
of humerus
Sternum
Shaft of humerus
Costal cartilage
Lateral epicondyle
Body of lumbar vertebra
Intervertebral disk
Iliac crest
Ilium
Pubis
Obturator
foramen
Patella
Tibial plateau
Tibial tuberosity
Tibia
Talus
Medial cuneiform
Intermediate cuneiform
Lateral cuneiform
Phalanges
Metatarsal bones

Coronal suture
Parietal bone
Lambdoid suture
Occipital bone
Temporal bone
Mastoid process
Atlas (C1)
Axis (C2)
Acromion
Spine of scapula
Scapula
Spinous
processes of
thoracic vertebrae
Floating ribs
(11–12)
Sacrum
Coccyx
Ischium
Head of femur
Shaft of femur
Femoral lateral condyle
Articular cartilage
Fibula
Lateral malleolus
Talus
Calcaneus
Cuboid   Navicular

Muscular system
Muscles of the upper body—front

THE MUSCULAR SYSTEM IS RESPONSIBLE FOR MOVEMENT OF THE HUMAN BODY. There are approximately 700 muscles attached to the bones of the skeletal system, and they constitute almost half of a person's body weight. The skeletal muscles support upright bipedal posture, generate movement, and provide a number of important functions, such as breathing and producing speech.

Skeletal muscles are named based on many different factors, including their anatomical position, such as the rectus abdominis of the abdomen; their origin and insertion point, such as the sternocleidomastoid, which connects the sternum and clavicle to the mastoid of the skull; the number of origins, such as a biceps or triceps; the shape, such as the deltoid or rhomboid; the size, by adding maximus or minimus to the end of the name, for example; and by the muscle's function, such as the supinator.

Muscles located on the anterior of the trunk are responsible for generating a number of different movements. The large and prominent pectoral muscles of the chest generate movement of the arm at the shoulder joint. The muscles of the anterior abdominal wall provide support for internal organs, play a role in breathing, and generate movements such as flexion, lateral flexion, and rotation of the trunk. While the rectus abdominus is the most recognizable, there are several muscles in this region that are organized in layers to allow for a variety of functions.

▶ **ABDOMINAL MUSCLES**
The abdominal muscles lie over a region not protected by the bones of the rib cage or the pelvic girdle. As such, these muscles play a critical role in protecting the delicate vital organs within the abdominal cavity.

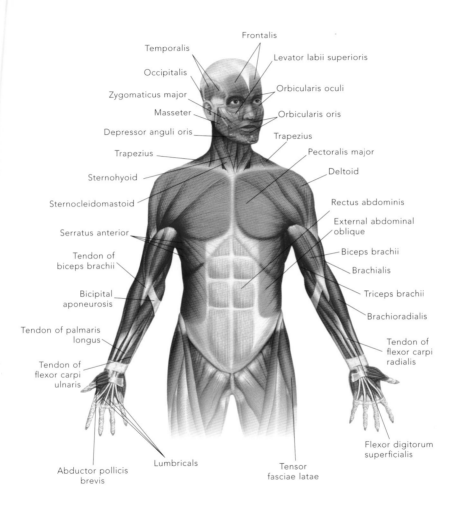

Frontalis

Temporalis

Levator labii superioris

Occipitalis

Zygomaticus major

Orbicularis oculi

Masseter

Orbicularis oris

Depressor anguli oris

Trapezius

Trapezius

Pectoralis major

Sternohyoid

Deltoid

Sternocleidomastoid

Rectus abdominis

External abdominal oblique

Serratus anterior

Tendon of biceps brachii

Biceps brachii

Brachialis

Bicipital aponeurosis

Triceps brachii

Brachioradialis

Tendon of palmaris longus

Tendon of flexor carpi radialis

Tendon of flexor carpi ulnaris

Flexor digitorum superficialis

Abductor pollicis brevis

Lumbricals

Tensor fasciae latae

Muscles of the lower body—front

THERE ARE A NUMBER OF MUSCLES AT THE FRONT OF THE HIP THAT CROSS THE HIP JOINT AND TETHER THE THIGH TO THE PELVIS. These muscles are primarily responsible for flexing and rotating the lower limb as well as playing an important role in standing posture and gait. The sartorius and rectus femoris are muscles that cross both the hip and knee joints and therefore play a role in the movement of each joint.

The musculature of the thigh can be split into three sections: anterior, medial, and posterior. In the anterior section, the quadriceps forms most of the muscle mass. The quadriceps group comprises four muscles: rectus femoris, vastus medialis (beneath rectus femoris), vastus lateralis, and vastus intermedius. These muscles contribute to the stability of the knee joint and are responsible for flexion of the knee during walking and running.

There are four muscles in the anterior compartment of the leg: tibialis anterior, extensor digitorum longus, extensor hallucis longus, and fibularis tertius. These muscles act to lift the foot so that the toes point upward and to roll the foot inward at the ankle joint. The extensor digitorum longus and extensor hallucis longus also extend the toes. These muscles are important during the swing phase of gait, in which they prevent the toes from dragging along the ground.

▶ **ADDUCTOR MUSCLES**

Located in the medial compartment, the adductor muscles (or groin muscles) move the thigh toward the body's midline. Included in this group are the adductor longus, adductor brevis, adductor magnus, pectineus, and gracilis muscles.

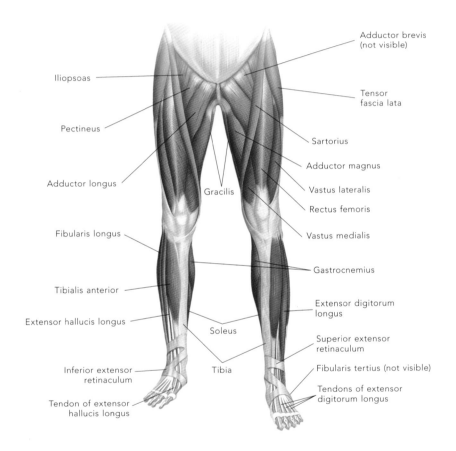

Adductor brevis (not visible)

Iliopsoas

Tensor fascia lata

Pectineus

Sartorius

Adductor magnus

Adductor longus

Vastus lateralis

Gracilis

Rectus femoris

Vastus medialis

Fibularis longus

Gastrocnemius

Tibialis anterior

Extensor digitorum longus

Extensor hallucis longus

Superior extensor retinaculum

Soleus

Fibularis tertius (not visible)

Inferior extensor retinaculum

Tibia

Tendons of extensor digitorum longus

Tendon of extensor hallucis longus

Muscular system—rear view

THE MUSCLES OF THE BACK STABILIZE AND SUPPORT THE VERTEBRAL COLUMN, while generating movements of extension and rotation of the spine. They can be divided into three groups: superficial, intermediate, and intrinsic. Superficial muscles are typically associated with movement of the shoulder; intrinsic muscles are associated with movement of the vertebral column. In addition, there are a number of muscles on the trunk that generate movement of the upper limbs. Some of these stabilize and move the scapula, while others support the shoulder joint and move the humerus.

Superficially, the trapezius covers the upper back, and its different fibers move the scapula up and down the trunk. The latissimus dorsi is located inferiorly to the trapezius and is responsible for extension and rotation of the arm. Deep to these muscles are a number of smaller muscles that act to stabilize and support the shoulder girdle, particularly through the upper limb's large range of motion. Situated along the spine are the paired erector spinae group that is attached to the

medial crest of the sacrum. These travel vertically, deep to the thoracolumbar fascia and within the groove of the vertebral column. They provide stability for bipedal posture as well as side-to-side rotation of the vertebral column.

The gluteal region is composed of powerful muscles that surround and stabilize the hip. These are divided into two groups: the deeper lateral rotators and the superficial abductors and extenders, which are situated superficially to the first group. This group includes the largest muscle in the body, the gluteus maximus, which, in conjunction with the hamstrings, extends the thigh through the hip joint. The three hamstring muscles, the biceps femoris, semitendinosus, and semimembranosus, form the posterior compartment of the thigh. In addition to extending the hip, the hamstrings are involved in flexion of the knee. Assisting with this is the superficial muscle, the gastrocnemius, which is located in the posterior compartment of the leg. Deep to the gastrocnemius is the soleus, and together they form the calf muscle.

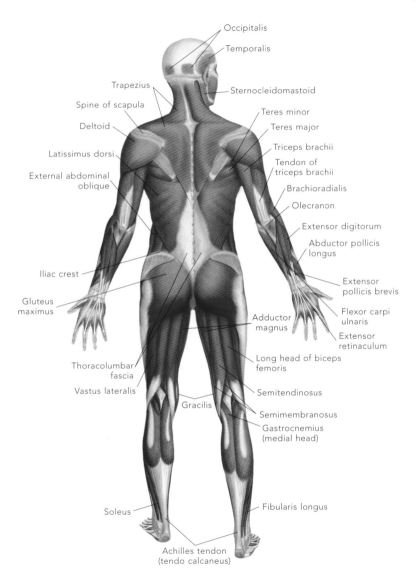

Occipitalis

Temporalis

Trapezius

Spine of scapula

Deltoid

Latissimus dorsi

External abdominal oblique

Iliac crest

Gluteus maximus

Thoracolumbar fascia

Vastus lateralis

Gracilis

Soleus

Sternocleidomastoid

Teres minor

Teres major

Triceps brachii

Tendon of triceps brachii

Brachioradialis

Olecranon

Extensor digitorum

Abductor pollicis longus

Extensor pollicis brevis

Flexor carpi ulnaris

Extensor retinaculum

Adductor magnus

Long head of biceps femoris

Semitendinosus

Semimembranosus

Gastrocnemius (medial head)

Fibularis longus

Achilles tendon (tendo calcaneus)

Muscular system—side view

THE DELTOID MUSCLE IS THE MOST PROMINENT MUSCLE THAT PROVIDES STABILITY TO THE SHOULDER JOINT. The deltoid has three sets of fibers lying anteriorly, posteriorly, and laterally. This allows the deltoid to abduct the arm (move it outward) and contributes to both flexion and extension of the arm.

The main flexors of the elbow joint are the biceps brachii and the brachialis, which lie on the anterior aspect of the arm. Although the biceps is located anteriorly to the humerus, they have no attachment to the humerus. Instead the biceps crosses the shoulder joint and contributes to shoulder flexion.

On the posterior aspect of the arm is the triceps brachii. This three-headed muscle is the only muscle on the posterior side of the arm. Crossing the shoulder joint, it contributes to shoulder extension and is the main extensor of the elbow joint.

In the forearm are the muscles that flex and extend the wrist. Many of the muscles that extend the wrist have a common origin point on the lateral epicondyle of the humerus, while many of the muscles that flex the wrist originate at the medial epicondyle. As such, these are common sites for overuse injuries such as tennis elbow and golfer's elbow.

In the lateral view, the fascia and its interconnected nature are clearly visible. Fascia is made up of collagen fibers that are irregularly arranged. It acts as a packing tissue and can withstand tensional forces in all directions.

▶ **FASCIA**

Fascia is an uninterrupted, three-dimensional web of tissue found throughout the body. In this illustration, the white thoracolumbar fascia can be seen along with the fascia lata extending down the leg. It functions to maintain structural integrity and to provide support and protection, and it acts as a shock absorber.

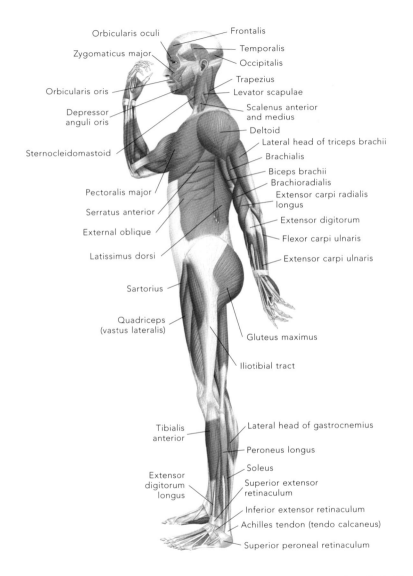

Orbicularis oculi
Zygomaticus major
Orbicularis oris
Depressor
anguli oris
Sternocleidomastoid
Pectoralis major
Serratus anterior
External oblique
Latissimus dorsi
Sartorius
Quadriceps
(vastus lateralis)

Frontalis
Temporalis
Occipitalis
Trapezius
Levator scapulae
Scalenus anterior
and medius
Deltoid
Lateral head of triceps brachii
Brachialis
Biceps brachii
Brachioradialis
Extensor carpi radialis
longus
Extensor digitorum
Flexor carpi ulnaris
Extensor carpi ulnaris

Gluteus maximus

Iliotibial tract

Tibialis
anterior

Extensor
digitorum
longus

Lateral head of gastrocnemius
Peroneus longus
Soleus
Superior extensor
retinaculum
Inferior extensor retinaculum
Achilles tendon (tendo calcaneus)
Superior peroneal retinaculum

Body regions

THE HUMAN BODY MAY BE DIVIDED AND SUBDIVIDED INTO A NUMBER OF REGIONS. The axial body comprises the head (cephalic region), neck (cervical region), and trunk, whereas the appendicular regions are made up of both the upper and lower limbs.

The cephalic region is subdivided into twelve regions, the most prominent including the posterior occipital region, frontal region (forehead), orbital region (eye), nasal region (nose), oral region (mouth), buccal region (cheek), and mental region (chin).

On the torso, the point of the shoulder is known as the acromial area. The chest in general is known as the thoracic region, while more specifically the sternum midline is the sternal region. The breast area is termed the mammary region, and the area under the armpit is the axillary region. Inferior to the sternum and superior to the pelvis is the abdomen, with the naval area referred to as the umbilical region.

On the upper limb, the upper arm is termed the brachial area, while the back of the elbow is the cubital and the front of the elbow is the antecubital. The forearm is the antebrachial, the wrist is the carpal region, and the palm of the hand is the palmar surface.

In the lower limb, the coxal area is around the hip, the buttocks are referred to as the gluteal region, and the inguinal region is the crease where the thigh meets the torso. The thigh itself is called the femoral region. The back of the knee is the popliteal, and the front of the knee is the patellar region. The leg is the crural region, but the outside of the leg is the fibular or peroneal region.

▶ **BODY REGIONS**

The body can be described in terms of regions, and these are helpful when referring to specific areas of the body, allowing more accurate clinical diagnosis of health problems.

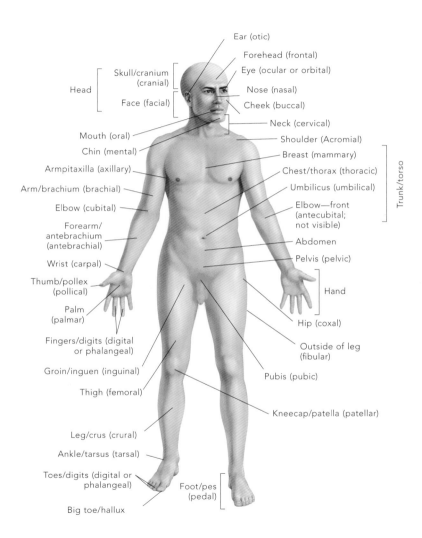

Ear (otic)

Forehead (frontal)

Eye (ocular or orbital)

Skull/cranium
(cranial)

Head

Nose (nasal)

Face (facial)

Cheek (buccal)

Neck (cervical)

Mouth (oral)

Shoulder (Acromial)

Chin (mental)

Breast (mammary)

Armpitaxilla (axillary)

Chest/thorax (thoracic)

Arm/brachium (brachial)

Umbilicus (umbilical)

Elbow (cubital)

Elbow—front
(antecubital;
not visible)

Forearm/
antebrachium
(antebrachial)

Abdomen

Wrist (carpal)

Pelvis (pelvic)

Thumb/pollex
(pollical)

Hand

Palm
(palmar)

Fingers/digits (digital
or phalangeal)

Hip (coxal)

Groin/inguen (inguinal)

Outside of leg
(fibular)

Thigh (femoral)

Pubis (pubic)

Kneecap/patella (patellar)

Leg/crus (crural)

Ankle/tarsus (tarsal)

Toes/digits (digital or
phalangeal)

Foot/pes
(pedal)

Big toe/hallux

Trunk/torso

Body movements

THE HUMAN BODY IS DESIGNED FOR
MOVEMENT, AND FOR EACH MOVEMENT
OF THE BODY THERE ARE ANATOMICALLY
CORRECT LABELS. All these terms
assume the body is starting in an
anatomically neutral position (known
as the "anatomical position"): an
upright posture, knees and arms fully
extended and next to the torso, with
the palms of the hands facing forward.

Flexion and extension are antagonistic
movements, flexion referring to the
movement that decreases the angle
between two body parts and extension
increasing that angle. For example,
flexion at the elbow decreases the
angle between the arm and the
forearm, whereas extension of the
elbow straightens the upper limb.
The ankle is an exception, as both
pointing the foot and raising the foot
are referred to as flexion: plantar flexion
and dorsiflexion, respectively.

The body movement of adduction
and abduction refers to the lateral
movement of a limb away from and
toward the midline of the body.
Abduction is movement away from the
anatomical position, and adduction is
returning it to the starting position.

Rotational movements are referred
to as pronation and supination. The
hands in the anatomical position are
supinated. Pronation of the forearm
forces the palms of the hand to rotate
until facing backward. In the leg,
rotational movements are referred to as
medial and lateral rotation, whereas the
circular rotation of the shoulder joint is
referred to as circumduction.

Finally, turning the foot inward so
that the sole of the foot faces the
midline is called inversion; turning it
out is eversion.

▶ **MOVEMENT TERMINOLOGY**
Familiarity with the correct terms to
describe movement facilitates the
description of the actions and functions
of muscles and ligaments, and also the
mechanisms of musculoskeletal injuries.

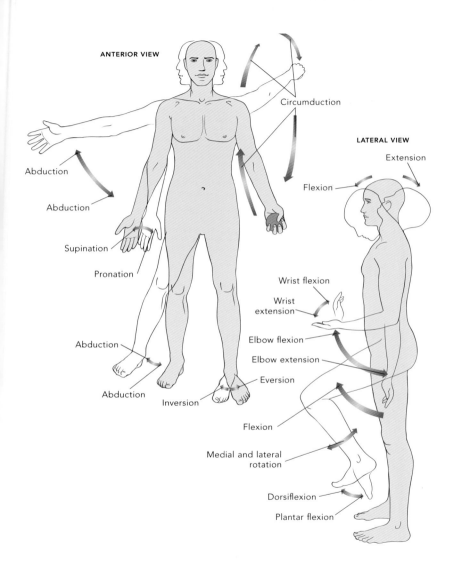

ANTERIOR VIEW

Circumduction

Abduction

Abduction

Supination

Pronation

Abduction

Abduction

Inversion

Eversion

LATERAL VIEW

Extension

Flexion

Wrist flexion

Wrist extension

Elbow flexion

Elbow extension

Flexion

Medial and lateral rotation

Dorsiflexion

Plantar flexion

Anatomical planes

TO FACILITATE THE UNDERSTANDING OF THE MOVEMENT OF MUSCLES, THE BODY CAN BE DIVIDED INTO TWO-DIMENSIONAL AREAS USING PLANES. Dividing the body into left and right halves is the sagittal, or median, plane, and all flexion and extension movements occur in this plane.

At a right angle to the sagittal plane is the coronal, or frontal, plane. All abduction and adduction from the anatomical position occur in the coronal plane.

Finally, a transverse, or horizontal, plane divides the body across from left to right, separating the upper body from the lower body.

TERMS COMMONLY USED TO DESCRIBE POSITIONS OF ANATOMICAL STRUCTURES:

Anterior: to the front of
Posterior: to the rear of
Inferior: below
Superior: above
Medial: toward the midline of the body
Lateral: away from the midline of the body
Proximal: closer to the trunk or a given reference point
Distal: away from the trunk
Superficial: close to the surface of the body
Deep: away from the surface of the body

▶ **DIVIDING THE BODY**
Medical imaging, such as X-rays, may be performed in any plane in order to get a clear picture of the part of the body being investigated.

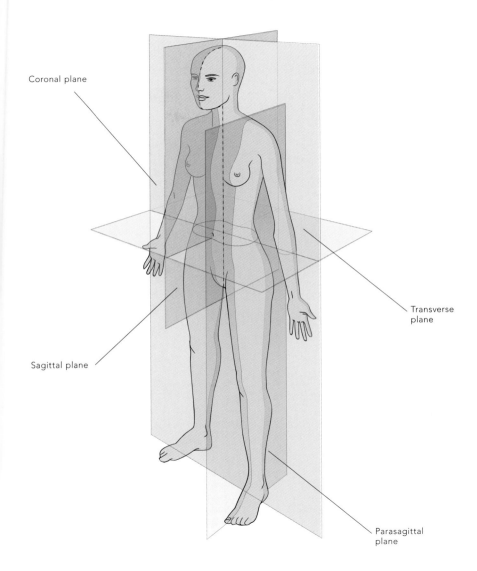

Coronal plane

Transverse plane

Sagittal plane

Parasagittal plane

Chapter 2:
Tissue types

The body consists of different tissue types, each of which has a role supporting, protecting, linking, and moving various body parts, limbs, and organs. In this section, we look at the remarkable properties of human tissue—its variability as well as its function. From tough rigid bone, to pliable tendons, skin, and muscle, tissue operates independently and together to provide the foundation upon which movement is based.

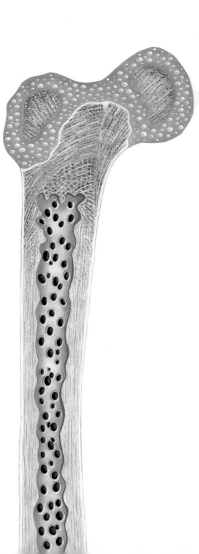

Bone

BONE IS A DENSE TISSUE DESIGNED TO
WITHSTAND AND TRANSMIT PHYSICAL
FORCES. Mineral salts create the
hardness of bone, while the matrix of
collagen fibers conveys strength. The
skeletal muscles are attached to the
bones via connective tissue and assist in
exerting forces on the bones to create
movement of the body. In addition to
providing support and movement,
some of the bones in the body have
bone marrow within them, which is
the site for the manufacture of new
blood cells.

Each bone in the body is classified
according to its shape (e.g., a short
bone, long bone, flat bone, or irregular
bone). When long bones, such as those
in the leg, are developing, they are
composed of three areas: a diaphysis,
an epiphysis, and an epiphyseal plate.
The diaphysis is the long shaft of the
bone and is composed mostly of hard
compact bone around a cavity
containing bone marrow.

The epiphysis is the end of the
bone, which has a lining of compact
bone around more cancellous, or

spongy, bone. The epiphyseal plate,
or growth plate, is the site of bone
growth where the bone lengthens
during development. In adulthood,
once the bones have reached full size,
the epiphyseal plate is replaced by the
epiphyseal line.

The outer layer of the bone is called
the periosteum and contains
connective tissue, blood vessels, and
nerves. It also provides the necessary
surface for anchoring tendons and
ligaments to the bone. The inner layer
of the periosteum contains bone cells
that are important for the bone's
growth and repair.

▸ **LIVING BONE**

Bones are far more complex than they
first appear. Compact bone forms a
hard outer layer, while the
honeycombed spongy bone gives the
bone strength in all directions. The
bone marrow in long bones also
performs hematopoiesis—the formation
of new blood cells from stem cells.

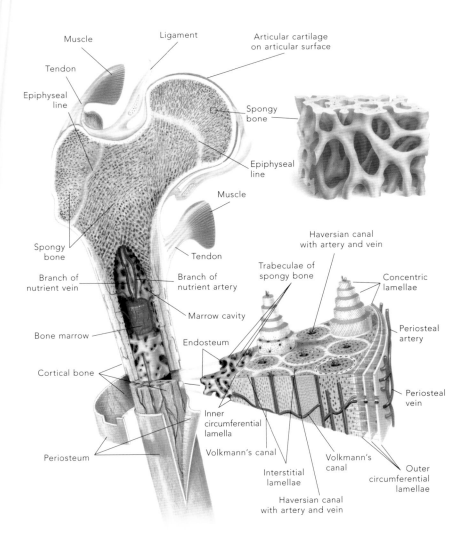

Muscle

Ligament

Articular cartilage
on articular surface

Tendon

Epiphyseal
line

Spongy
bone

Epiphyseal
line

Muscle

Haversian canal
with artery and vein

Tendon

Trabeculae of
spongy bone

Concentric
lamellae

Spongy
bone

Branch of
nutrient vein

Branch of
nutrient artery

Periosteal
artery

Marrow cavity

Bone marrow

Endosteum

Periosteal
vein

Cortical bone

Inner
circumferential
lamella

Volkmann's
canal

Periosteum

Volkmann's
canal

Interstitial
lamellae

Outer
circumferential
lamellae

Haversian canal
with artery and vein

Bone growth

BONES DEVELOP EITHER DIRECTLY IN THE EMBRYONIC STATE—by the deposition of mineral salts—or in a preformed cartilage model. The cartilage template of long bone is laid down during fetal development. The cartilage gradually hardens, or ossifies, with ossification eventually extending almost to the end of the bone. Growth then begins at both ends of the bone, with the cartilaginous epiphyseal plate forming the junction between the existing bone and new bone growth. When fully ossified, growth in bone length is complete.

Every tissue in the human body is replaced and regenerated continuously. With some cells this is a simple process: dead skin cells fall off and are replaced with new cells underneath, red blood cells die and are broken down to be replaced by new cells. As bone tissue is hard and mineralized, the process of remodeling requires a few more cells to play a role.

Osteoblasts form new bone through osteogenesis. The osteoblasts produce many cell products, including enzymes, growth factors, hormones, and collagen. Eventually, the osteoblast is surrounded by the growing bone matrix and the cell becomes trapped in a space called a lacuna as the material calcifies. The osteoblast then becomes an osteocyte, or bone cell.

Osteoclast cells remove bone tissue, breaking it down by producing an acidic environment and decalcifying the bone matrix and producing enzymes to digest the collagen components. Bone remodeling is dependent on the balanced activity of osteoblasts and osteoclasts.

▶ **COMPACT BONE**
Compact bone is organized into units called osteons. Each osteon is arranged around a central canal that houses an artery, carrying the bone's blood supply.

BONE GROWTH

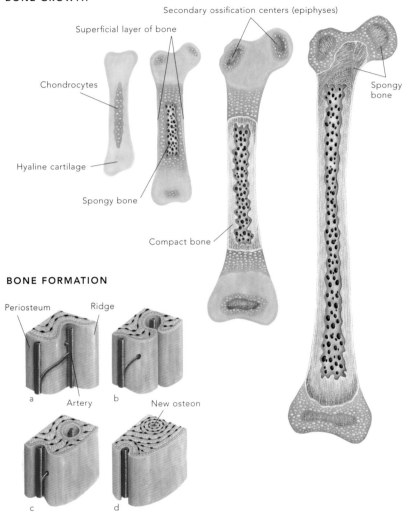

Secondary ossification centers (epiphyses)

Superficial layer of bone

Chondrocytes

Spongy bone

Hyaline cartilage

Spongy bone

Compact bone

BONE FORMATION

Periosteum   Ridge

a   Artery

b

New osteon

c   d

Muscles

MUSCLES ARE DESIGNED FOR MOVEMENT. Through contraction they move substances around the body and exert mechanical forces on bones to move the body itself.

There are three types of muscle tissue in the body: smooth muscle, cardiac muscle, and skeletal muscle.

Smooth muscle is found in the linings of blood vessel walls, where it helps to regulate blood flow through the vessel. It is also found in the walls of the gastrointestinal tract (e.g., the muscles that control peristalsis) and in the eye (e.g., the muscles that control the shape of the eye lens). These muscles undergo slow and sustained contractions that pull in every direction. We don't have conscious control of this muscle tissue, so it is known as an involuntary movement.

Cardiac muscle is located in the heart wall and is characterized by its endurance and high contractile strength. Coordinated contraction of the cardiac muscle is highly organized, thus allowing the heart to efficiently pump blood around the body. The cardiac muscle is also involuntary; however, it differs from smooth muscle in that it has a striated appearance.

Skeletal muscles are composed of multiple bundles of muscle fibers, which are composed of actin and myosin filaments. These filaments are arranged in a repeating unit known as a sarcomere and give skeletal muscle its striated appearance. The sarcomeres are for contraction and the production of a physical force in the direction of the fibers. This muscle is termed skeletal muscle since it always attaches to the skeleton to exert force on bones. Each muscle has an origin and insertion point where it attaches to the skeleton at each end. Skeletal muscle is known as voluntary muscle, as we usually have conscious control over it.

▶ **MUSCLE TISSUE**
The types of muscle tissues differ not only in their appearance, but also in the way they function.

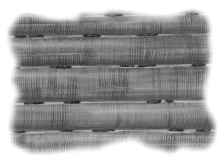

Skeletal muscle

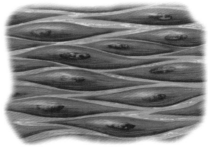

Smooth muscle

Cardiac muscle

Cartilage

CARTILAGE IS A TYPE OF HARD BUT
FLEXIBLE CONNECTIVE TISSUE
COMPOSED MAINLY OF WATER AND A
DENSELY PACKED MATRIX OF COLLAGEN
AND ELASTIC FIBERS. Specialized cells
known as chondrocytes produce this
cartilaginous matrix, which
subsequently traps the cell and
prevents migration.

Cartilage, unlike other connective
tissue, has no blood vessels or nerves.
As such, damaged cartilage has a very
limited ability to heal.

There are three types of cartilage
in the body: elastic cartilage,
fibrocartilage, and hyaline cartilage.

Elastic cartilage is the most flexible
of the three types and is able to
withstand repeated deformation. It is
only found in the ear and throat and
provides a greater degree of flexibility
while still offering the necessary
support to maintain its shape.

Fibrocartilage is very strong and
compressible and is found in the body
where there are highly compressive
and tensile forces. Fibrocartilage is
found in the intervertebral discs as
well as in joints that require only a
limited range of movement, such as
the pubic symphysis.

Hyaline cartilage is the weakest form
of cartilage and forms a 2–4 mm
covering on the ends of bones in most
movable joints. It provides a smooth
surface for articulation with the other
bones in the joint. Movement of water
in and out of the matrix gives this
cartilage the ability to cushion against
loads in the joint. Hyaline cartilage is
also used as scaffolding in the airways.

▶ **HYALINE CARTILAGE**

This cartilage, found at the end of
bones, has a shiny, glassy appearance.
This surface maintains smooth
movement of the joints. Under the
microscope, the chondrocytes are
visible in their matrix.

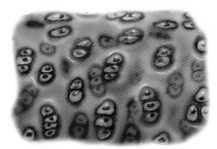

Hyaline cartilage

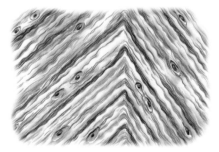

Fibrocartilage

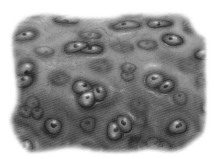

Elastic cartilage

Tendons and ligaments

TENDONS AND LIGAMENTS ARE
FIBROUS CONNECTIVE TISSUES THAT
TETHER BONE TO BONE AND MUSCLE
TO BONE, RESPECTIVELY.

A ligament is composed of dense
fibrous bundles of collagen and cells
known as fibrocytes. In some joints,
ligaments form a capsular sac that
encloses the joint and creates an
environment in which the joint
remains lubricated. The ligaments
prevent excessive movement that could
damage the joint. Typically, the more
ligaments a joint has, and the less
movement they allow, the more stable
the joint. However, this stability may
mean reduced mobility.

Tendons are flexible, elongated
bundles of fibrous collagen that
connect a muscle to a bone. Tendons
are highly resistant to extension, and
when subjected to mechanical stress,
such as during the transmission of
muscular contractile force, they
enable body movement. The
musculotendinous junction is the area
between the muscle and its tendon and
is considered the growth plate of the
muscle. The collagen fibers of the bone
periosteum merge with those of the
tendon to allow forces to be
transmitted from the contracting
muscle to the bone. Elastic properties
of the tendon itself can save energy by
functioning as a spring during
movements such as walking.

▶ **STRUCTURE**

Both ligaments and tendons are
arranged in bundles, which are layered
parallell to each other. They align
perpendicularly to the direction of the
forces they are exposed to.

Ligament tissue (relaxed)

Tendon tissue

Tendon tissue (relaxed)

Ligament tissue

Neural tissue

NEURAL TISSUE IS RESPONSIBLE FOR
TRANSMITTING SIGNALS RAPIDLY FROM
ONE PART OF THE BODY TO ANOTHER.
The nervous system comprises the
central nervous system (CNS) and the
peripheral nervous system (PNS). The
CNS is composed of the brain and
spinal cord, which control the body,
while the PNS is made up of all the
other nerves that transport sensory and
motor information to and from the
body and CNS.

The nervous system is composed
of two different types of cells: neurons
and neuroglia. Neurons transmit
electrochemical signals and have a cell
body that contains a nucleus and other
organelles, while branch-like structures
called dendrites communicate with
surrounding neurons. Neurons also
have a long axon that allows the
sending of signals to other neurons or
tissues. The axon terminal is the most
distal part of the axon, where the
neuron forms a junction where
neurotransmitters are released.
Neuroglia, or glial cells, play an
insulating and protective role for
neurons. These cells surround and
protect neurons, which helps improve
the transmission of signals and
maintains a healthy nervous system.

There are three types of neurons:
afferent, or sensory, neurons that
transmit sensory information from
receptors (such as pain receptors)
back to the CNS; efferent, or motor,
neurons that transmit signals from
the CNS to effectors in the body,
such as muscles; and interneurons
that work as a way of integrating
information from afferent neurons
to direct efferent neurons.

▶ **NEURAL TISSUE**
98% of neural tissue is located in
the brain and spinal cord. This central
nervous tissue is the control center
for the whole nervous system.

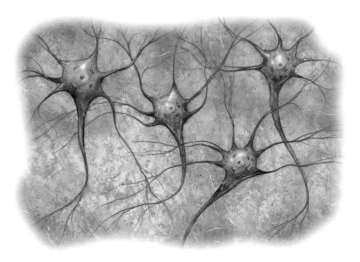

Neural tissue

Skin

SKIN COVERS THE ENTIRE EXTERNAL SURFACE OF THE HUMAN BODY AND ACTS AS A PROTECTIVE BARRIER TO prevent damage to internal tissues from trauma, ultraviolet radiation, excessive light, temperature extremes, toxins, and bacteria. It also houses sensory receptors, such as those for pain, heat, cold, and touch, and plays an important role in regulation of body temperature. Skin is composed primarily of two layers, the epidermis and dermis, which cover a third, fatty layer.

The epidermis is the outer layer of epithelial cells which are supported by the deeper dermis. It is primarily made up of cells called keratinocytes. They originate from cells in the deepest layer of the epidermis called the basal layer and slowly migrate up toward the surface of the epidermis. Once the keratinocytes reach the surface, they are gradually lost (through shedding) and are replaced by newer cells making their way up from below. Throughout the basal layer of the epidermis are cells called melanocytes, which produce the pigment melanin, giving skin its color.

The epidermis also contains Langerhans cells, which are part of the skin's immune system.

Deep to the epidermis is the dermis, a thick layer of fibrous and elastic tissue that gives the skin its flexibility and strength. It contains nerve endings, sweat glands and sebaceous glands, hair follicles, and blood vessels that supply nutrients to all skin layers.

With their base coiled deep in the dermis, sweat glands extend through the layers to the surface. Eccrine sweat glands secrete sweat, a mixture of water, salts, and fats. In warm conditions with low humidity, perspiration (secretion of sweat) and evaporation cool the body. Apocrine sweat glands, which become active at puberty, are larger and deeper, and they produce thicker secretions than eccrine glands. They are primarily found in the armpit and around the genitals. Rather than opening directly onto the surface of the skin, apocrine glands secrete sweat into the pilary canal of the hair follicle and are most active in times of stress and sexual arousal.

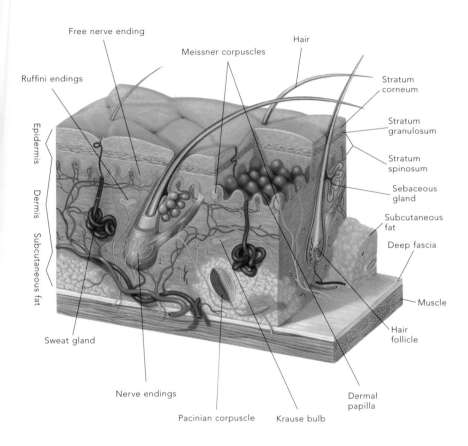

Free nerve ending

Meissner corpuscles

Hair

Ruffini endings

Stratum corneum

Stratum granulosum

Stratum spinosum

Sebaceous gland

Subcutaneous fat

Deep fascia

Muscle

Hair follicle

Dermal papilla

Sweat gland

Nerve endings

Pacinian corpuscle

Krause bulb

Epidermis

Dermis

Subcutaneous fat

▲ HAIR

Similar to skin, hair is primarily composed of keratin proteins. The hard, ropelike protein filaments provide structure and strength to the hair shaft.

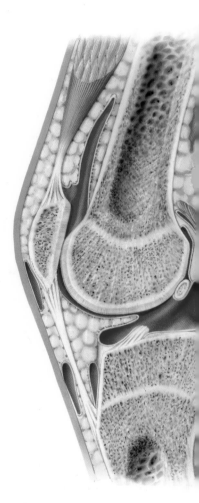

Chapter 3:
Joints

The inflexible nature of bone means that movement only occurs at the joints where they meet. In this chapter, we look at how joints determine the nature and range of a movement, and also act as a safety mechanism to ensure that movements are not excessive. Or, as in some regions such as the skull, how they act to provide a rigid, protective structure where movement is minimal.

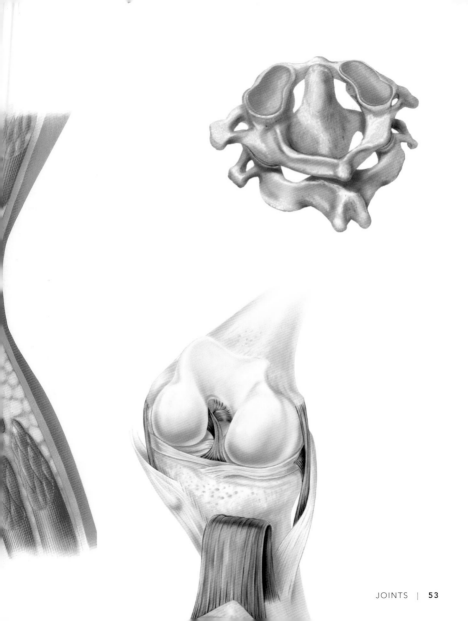

Introduction to joints
Types of joints

SYNOVIAL JOINTS ARE THE MOST COMMON—AND THE MOST COMPLEX—JOINTS IN THE BODY. In the joint space is a lubricant known as synovial fluid. Synovial fluid moves around the cavity of a joint to provide lubrication, distribute nutrients, and absorb impact forces. Since cartilage has a very poor blood supply, the synovial fluid's ability to deliver nutrition is vital for maintaining healthy cartilage.

Enclosing the joint space is the articular capsule, a fibrous ligament structure that is continuous with the periosteum of the bone. The outer layer—or capsular ligament—provides support to the joint, while the inner layer—the synovium or synovial membrane—is highly vascularized and regulates the exchange of nutrients.

Around the joint there are usually accessory ligaments that are separate from the joint capsule. These bundles of dense, regular connective tissue prevent any extreme movements that may damage the joint. In addition, crossing the joint from one bone to another are muscles and their tendons that move the joint and act to stabilize it. Around the joint, particularly underneath the tendons where friction will occur, are bursae, which are small sacs filled with synovial fluid, enabling the joint to have free movement.

▶ **THE ANATOMY OF A SYNOVIAL JOINT**

Illustration showing the component parts of a synovial joint. Sudden impact or unusual force to a joint may move the bones out of their normal position. If the bones are moved out of position so that their articular surfaces are no longer in contact, this is called a dislocation.

SYNOVIAL JOINT

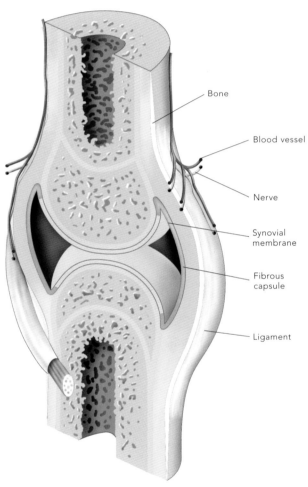

Bone

Blood vessel

Nerve

Synovial membrane

Fibrous capsule

Ligament

Types of joints *(cont.)*

Cartilaginous joints connect bone with cartilage and are classified as either primary or secondary joints. In a primary cartilaginous joint—known as a synchondrosis—the joint offers little movement and is predominantly composed of hyaline cartilage. An example of a synchondrosis joint is the articulation between the sternum and the first rib.

Secondary cartilaginous joints—or symphyses—offer a small degree of movement and are composed of both hyaline cartilage and fibrocartilage. An example of a symphysis joint is the pubic symphysis. Cartilaginous joints provide a high level of stability while still maintaining the ability to absorb shock.

Fibrous joints connect bones by means of a tough fibrous tissue and provide little to no movement. There are three types of fibrous joints: sutures, gomphoses, and syndesmoses. Sutures are immovable joints such as those found between the flat bones of the skull. Gomphoses, which are also immovable, are found in the periodontal ligament between the teeth and the mandible and maxilla. Syndesmoses are joints held together by a type of flat ligament called an interosseous membrane. This binds two bones together along their length. An example of this is the connection between the tibia and fibula.

▶ **SHOCK ABSORPTION**

The hyaline cartilage between sternum and rib (opposite, above left) is an example of a primary cartilaginous joint. The interosseous membrane between tibia and fibula (far right) is an example of a syndesmosis. Both types provide support and offer limited movement.

RIB CAGE ANTERIOR VIEW

TIBIA AND FIBULA

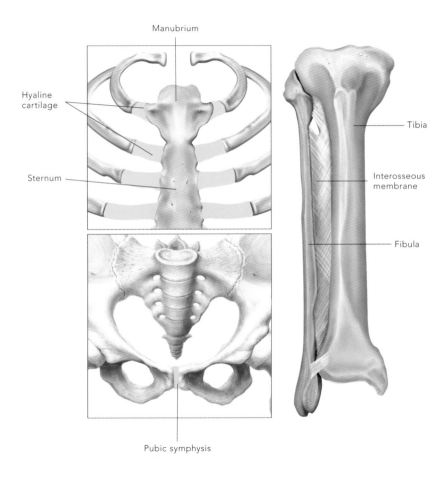

Manubrium

Hyaline cartilage

Sternum

Pubic symphysis

Tibia

Interosseous membrane

Fibula

Types of synovial joints

THERE ARE SIX DIFFERENT TYPES OF
SYNOVIAL JOINTS IN THE BODY THAT
PROVIDE A RANGE OF MOBILITY.

Gliding, or plane, joints are the most
common type of synovial joint and are
formed between bones that meet at
two flat articular surfaces. Gliding
joints allow the bones to slide past one
another in any direction along the flat
surface of the joint. The angle of the
bones relative to each other does not
change, and only very limited rotation
occurs. The joints between the carpal
bones in the hand and the tarsal bones
in the foot are gliding joints.

Ellipsoidal, or condyloid, joints have
an oval-shaped process that fits into an
elongated or ellipsoidal cavity of the
other bone. This joint allows
movement in two directions—flexion
and extension or abduction and
adduction—or a combination of the
two, which is circumduction.

A saddle joint is a synovial joint in
which one of the bones forming the
joint is shaped like a saddle. The other
bone sits across the joint like the legs of
a rider on a horse. Saddle joints

provide more flexibility than a hinge or
gliding joint, as they can move in an
oval range of motion in all directions,
similar to an ellipsoidal joint.

▶ **THE HAND AND WRIST**
The joints shown right make up the
hand and wrist. The configuration of
the carpal bones creates gliding joints
between each of them. In combination
with the ellipsoidal joint of the wrist
and the saddle joint at the base of the
thumb, it allows great mobililty across
multiple planes.

GLIDING JOINT

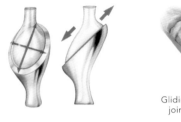

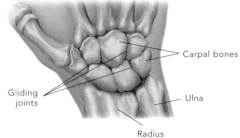

Carpal bones

Gliding joints

Ulna

Radius

ELLIPSOIDAL JOINT

Scaphoid bone

Ellipsoidal joint

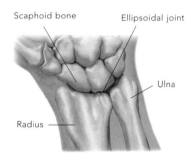

Ulna

Radius

SADDLE JOINT

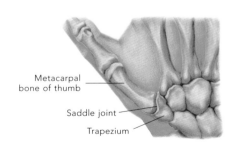

Metacarpal bone of thumb

Saddle joint

Trapezium

Types of synovial joints *(cont.)*

The pivot joint has a limited range of movement in which one bone moves around the axis of another, producing a rotational movement.

A number of joints in the body are classified as hinge joints. The shape of the joint usually facilitates a large range of movement, but the shape of the articular surfaces restricts this to one axis or direction. As such, hinge joints only allow flexion and extension. Some hinge joints allow a small degree of rotation or lateral movement, but this is restricted by the joint structures.

Ball-and-socket joints allow the greatest range of movement and mobility of all of the synovial joints. The only examples of this joint in the body are those at the shoulder and the hip. Ball-and-socket joints allow for movement in all planes as well as circumduction, to a greater extent than other synovial joint types such as ellipsoidal and saddle joints.

The ligament structures of these joints need to be less tight than in other joints to allow this movement; therefore, stability and movement must be maintained by a great deal of musculature.

▶ **THE ELBOW AND HIP JOINTS**

The elbow joint is both a hinge and a pivot joint. The articulation between the radius and ulna forms the pivot, while the radius, ulna, and humerus together form the hinge. The hip joint is necessarily more stable than the shoulder joint, because it helps transfer weight from the upper body to the lower limbs.

PIVOT JOINT

Axis and atlas

 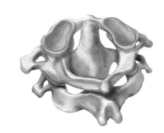

HINGE JOINT

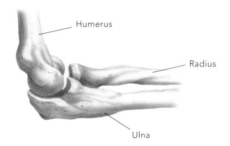

Humerus

Radius

Ulna

BALL-AND-SOCKET JOINT

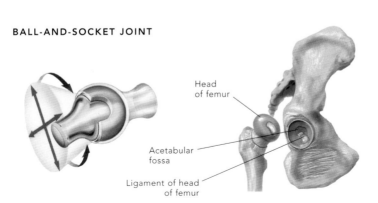

Head
of femur

Acetabular
fossa

Ligament of head
of femur

Detailed structure of a synovial joint

IN THE SYNOVIAL JOINT, THE SYNOVIUM IS THE SOFT TISSUE THAT LINES THE SPACES OF THE JOINTS, TENDON SHEATHS, AND BURSAE. The synovium covers the entire internal space of the joint, except for the articular surface that is lined with hyaline cartilage. It consists of a continuous surface layer of cells, called an intima, and an underlying tissue, called the subintima. The subintima consists of intra-articular vessels, such as blood vessels and lymphatic vessels, and nerves. The cells that produce the synovial fluid, synoviocytes, also reabsorb the fluid in a constant recycling process.

The synovial fluid itself is translucent and pale yellow in color. Synovial fluid has two main functions: to aid in the nutrition of articular cartilage by acting as a transport medium for nutrients such as glucose and oxygen, and to smooth the mechanical function of joints by lubricating the articulating surfaces. As this fluid fills the joint space, it also plays a role in shock absorption when the joint comes under load. Since cartilage has no blood supply of its own, it must get its entire nutrient supply via the synovium.

Hyaline cartilage in the synovial joint has four zones: the superficial zone, the middle zone, the deep zone, and the calcified zone. The most superficial zone is tightly packed with collagen fibers that run parallel to the direction of movement of the joint, providing the hard, smooth surface. The intermediate and deep zones provide resistance to compressive forces, giving cartilage its shock-absorption qualities. The calcified layer secures the cartilage to bone by anchoring the collagen fibers of the deep zone to subchondral bone.

▶ **STRUCTURE OF THE KNEE JOINT**
The knee joint is a bicondylar synovial joint, formed by articulations between the patella, femur, and tibia. The shape of the knee joint means that it is relatively weak, and so it relies on muscles and ligaments to maintain stability.

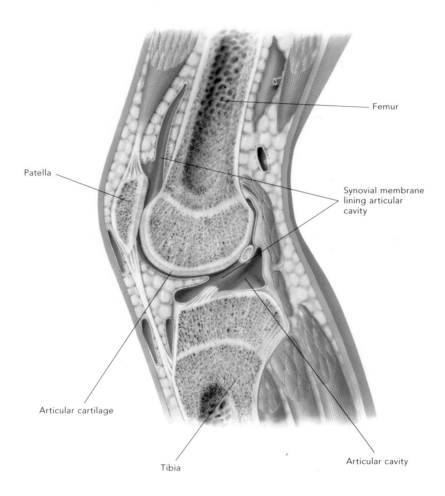

Femur

Patella

Synovial membrane lining articular cavity

Articular cartilage

Tibia

Articular cavity

Ligaments

LIGAMENTS ARE DENSE BANDS OF COLLAGEN TISSUE THAT ARE ALIGNED IN PARALLEL AS INTERCONNECTED FIBER BUNDLES. They vary in size, shape, and orientation, which is of critical importance with regard to the type of joint each ligament is associated with.

The outer layer of the ligament is the epiligament and is a vascularized layer that contains sensory and proprioceptive nerves. These nerves travel in close proximity to the blood vessels and provide the central nervous system with information about the angle of the joint.

Though the ligament appears as a single structure, with joint movement some fibers appear to tighten or loosen depending on the bone positions. While not under load, there is a crimping pattern in the ligament structure. This is where the collagen fibers are cross-connected to each other—running straight and parallel but also showing waviness or bends in the fibers.

This allows the ligament to recoil while not under load and to provide some slack in the structure, where under load the ligament will not be stiff in the initial phase while "un-crimping" or straightening out.

▶ **LIGAMENT**
The structural properties of a ligament in a relaxed state and under tension.

Ligament in relaxed state

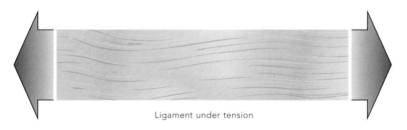

Ligament under tension

Ligaments of the knee joint

THE FUNCTION OF LIGAMENTS IS TO PROVIDE STABILITY TO A JOINT WHILE MAINTAINING MOBILITY. A good example is the knee joint and its associated ligaments. The knee has to remain stable to support the weight of the body and forces generated during running and jumping. Several muscles cross the knee joint to contribute to the stability of the knee; however, it is the system of ligaments that limits any unwanted movement. The joint itself is quite shallow, so forces could easily move the femur off the condyles of the tibia without the ligament structure bolstering the joint.

The medial view of the knee shows the tibial collateral ligament, a long, flat band of fibers that extend from the femoral condyle to the shaft of the tibia. This ligament is fused to the capsule and the medial meniscus. If the knee is struck with force, the medial collateral ligament resists valgus deformation. This is movement of a bone or joint that is twisted away from the body.

The fibular collateral ligament reinforces the lateral side and is a narrow band of fibers extending from the lateral femoral condyle down to the head of the fibula. Unlike the medial collateral ligament, it is not attached to the knee capsule or lateral meniscus and is therefore less susceptible to injury. This ligament resists varus deformation, which is the inward movement of the distal part of the leg.

The anterior and posterior cruciate ligaments lie within the joint capsule. The anterior cruciate ligament passes from the front of the tibia up and posteriorly to the rear of the femur. The posterior cruciate ligament passes from the back of the tibia up and anteriorly to the front of the femur. In this position they form a cross shape, giving them their name. The anterior cruciate ligament prevents excessive forward gliding of the tibia, while the posterior cruciate prevents excessive backward gliding of the tibia.

The combination of these ligaments ensures that the movement of the knee is limited primarily to flexion and extension, allowing for efficient muscle function and protection of the structures around the joint.

▶ KNEE LIGAMENTS

As they play an important role in stability, the knee ligaments can be exposed to high forces during sporting activities. As such, they may be sprained (overstretched), or sometimes ruptured (torn). A rupture may be partial, involving just some of the fibers, or complete, where the ligament is torn through completely.

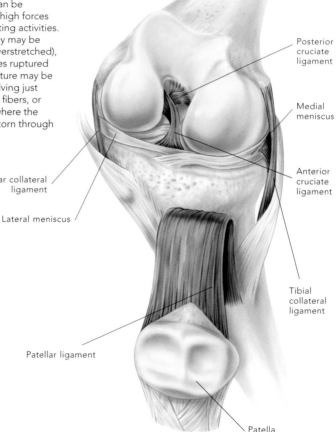

Femur

Posterior cruciate ligament

Medial meniscus

Anterior cruciate ligament

Fibular collateral ligament

Lateral meniscus

Tibial collateral ligament

Patellar ligament

Patella

Ligament properties

WHEN A LOAD IS APPLIED TO A
LIGAMENT OR TENDON, IT WILL
LENGTHEN DUE TO ITS INHERENT
ELASTIC PROPERTIES. Initially, it will
lengthen easily due to the crimped
collagen fibers. Continued loading will
result in increasing stiffness and length
until a stage is reached where it exhibits
complete linear stiffness. If the load is
further increased, the ligament will
continue to absorb energy until it
eventually ruptures.

Viscoelasticity is a property of
ligaments that allows them to be
viscous and elastic. In ligaments this is
known as creep and stress relaxation.
Creeping is the lengthening of the
ligament under constant load, whereas
stress relaxation is the dissipation of
tension if the ligament is held at the
same length for a prolonged period
of time.

▶ **STABILITY**
The ligaments of the interphalangeal
joints, or the joints on the fingers, are
stretched on extension. This provides
stability to the finger. These are often
injured in sports by a blow to the
straightened finger.

▶ **LAXITY**
Ligaments cannot remain tense in all
joint positions. When relaxed, the
ligament is unable to provide support
or stability to the joint.

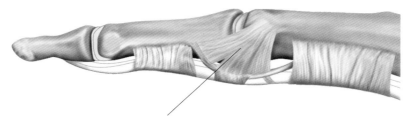

Collateral ligaments stretched

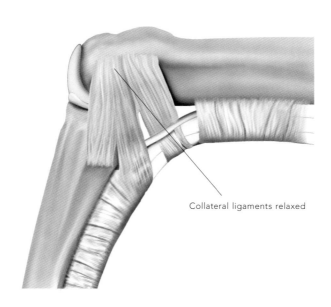

Collateral ligaments relaxed

Range of movement

RANGE OF MOVEMENT, OR RANGE OF MOTION, DESCRIBES THE FREEDOM WITH WHICH A JOINT CAN MOVE. In a simple joint—such as the hinge joint of the knee—the range of motion is measured in degrees of flexion and extension. In a joint with movement in different planes—such as the shoulder joint—the range of motion is measured in either degrees of flexion and extension, adduction and abduction, or internal and external rotation. A certain range of motion at each joint is required for the activities of daily living, but a greater range may be required by the demands of certain sporting activities.

There are two types of range of motion: active and passive. In active range of motion, the end of range is as far as the joint can move through the use of muscles alone around that joint. While passive range of motion is greater, external forces such as stretching against an immovable object are required to assist.

Range of motion around a joint may be limited by a number of factors. In some cases, the degree to which a joint can move is limited by the stiffness of the muscles around that joint. In other cases, there might be a bony endpoint, where two bones come together and limit further changes in joint angle. In many instances, the ligaments supporting the joint will limit the degree to which it will move. In some joints, soft tissue such as muscle and fat mass may reduce the ability of a joint to flex beyond a certain point.

▶ **LIMITED RANGE OF MOTION**
Range may be limited by injury or diseases such as osteoarthritis. Pain, swelling, and stiffness can limit the range of motion and also impair function and the ability to perform daily activities. After injury or surgery, restoring range of motion is one of the first goals of rehabilitation.

FLEXION OF THE ARM

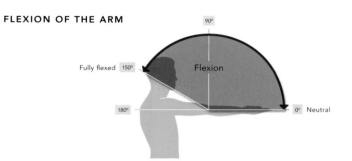

90°

Fully flexed 150°

Flexion

180°

0° Neutral

MECHANICS OF THE KNEE

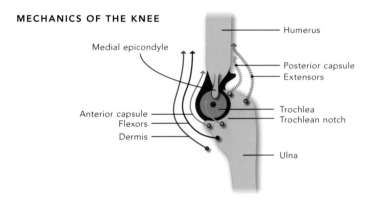

Humerus

Medial epicondyle

Posterior capsule
Extensors

Anterior capsule
Flexors
Dermis

Trochlea
Trochlean notch

Ulna

KNEE FLEXED

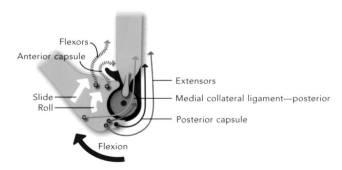

Flexors
Anterior capsule

Extensors

Slide
Roll

Medial collateral ligament—posterior

Posterior capsule

Flexion

Joint levers

WHILE MUSCLES CONTRACT TO GENERATE FORCE AND CREATE MOVEMENT, THE BONES AND JOINTS CREATE LEVERS TO FACILITATE THIS MOVEMENT. In a biomechanical context, a lever is described as being composed of a fulcrum, a rigid beam, and the force required to overcome a resistance or load. In the body, the muscle provides the force, the joint acts as the fulcrum, and the weight of the limb or body is the load.

Levers can be used to magnify movement; for example, when kicking a ball, small contractions of leg muscles produce a much larger movement at the end of the leg. Or, levers can be used to give a strength advantage to lift something with less effort. Different joints have different levers to elicit these two advantages, but one joint can only have one advantage or the other, not both.

There are three classes of levers in the human body. A first-class lever has the fulcrum directly between the muscle force and the load, like a seesaw in a playground, where two opposing forces balance on either side of the fulcrum.

In a second-class lever, the load is between the fulcrum and the muscle force, such as in a wheelbarrow. This lever creates a mechanical advantage so that the muscle force required is less than the load force.

A third-class lever is where the load is farther away from the fulcrum than the muscle force. An example of this is using a canoe paddle, where the upper hand stays stationary (fulcrum) and the lower hand pulls on the paddle to overcome the resistance of the water. There is no mechanical advantage here. In fact, the effort is greater than the load; however, this disadvantage is compensated with a larger movement.

▶ **LEVERS IN THE BODY**
Right (from the top): first-, second- and third-class levers. While examples of all three levers are found in the body, most levers are third class and thus convey an advantage of generating larger movements.

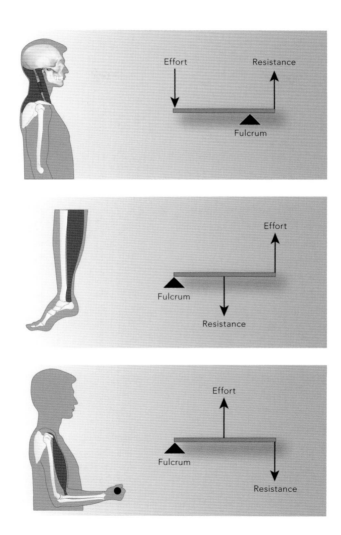

Effort

Resistance

Fulcrum

Fulcrum

Effort

Resistance

Fulcrum

Effort

Resistance

Range of movement injuries

IN THE ELBOW JOINT, EXTENSION IS
NORMALLY LIMITED BY THE OLECRANON
PROCESS, WHICH COMES INTO CONTACT
WITH THE OLECRANON FOSSA.
However, if an individual sustains an
injury to the tendon of the biceps
brachii, this will result in swelling,
pain, and inflammation that may limit
extension before this point.

Some athletes require a greater
range of motion than usual in order
to perform the movements required in
their sport. However, excessive range of
motion may increase the risk of injury.
One example is where the individual
has sustained a ligament injury in the
past and, although healed, the ligament
may not support the joint or allow
for a normal range of movement.
This situation puts the individual at
a higher risk of reinjury and increases
the likelihood of developing
osteoarthritis due to increased stress
in the articular cartilage.

▶ **HYPERMOBILITY**

Joint hypermobility is commonly
associated with joint injuries. Athletes
with hypermobility have a significantly
increased risk of knee joint injury during
sporting activities that involve contact
or collisions.

HYPEREXTENSION

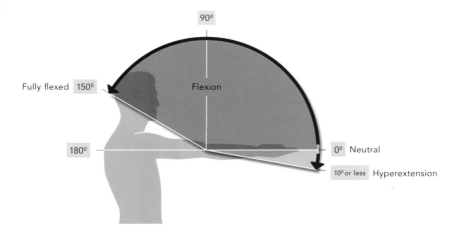

90⁰

Fully flexed 150⁰

Flexion

180⁰

0⁰ Neutral

10⁰ or less Hyperextension

Chapter 4:
Skeletal muscle

In this chapter, we look at how skeletal muscle has evolved as a remarkable structure that fuels and facilitates human movement. Reacting to nervous stimuli, the myriad of fibers and filaments create contractions and forces that determine the speed, range, and strength of movement that an individual produces. Skeletal muscle is the body's engine room, continuously producing the energy and movement that are essential for life.

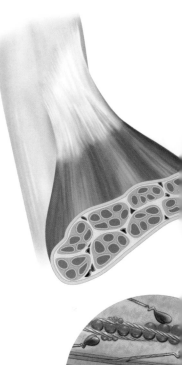

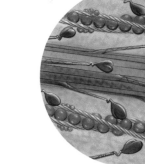

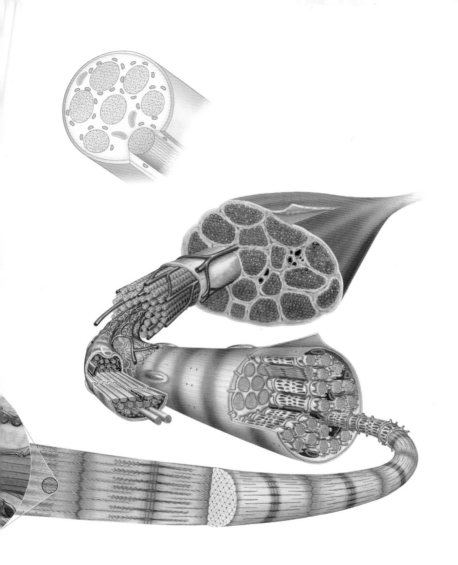

Muscle tissue
Muscle tissue types

A SKELETAL MUSCLE IS A MUSCLE THAT ALWAYS HAS ATTACHMENTS TO THE BONES OF THE SKELETON. When skeletal muscles contract, they shorten and create tension that pulls the two attachment sites closer together. Skeletal muscle has its fibers arranged so that they run parallel to each other, giving it a striated appearance. This striated appearance is also found in cardiac muscle—the muscle of the heart wall that contracts to squeeze blood from the heart into the systemic and pulmonary circulation. Like skeletal muscle, the heart muscle also has highly organized fibers that run parallel to each other. However, heart muscle fibers also branch out and interconnect via anchoring sites called intercalated discs.

▶ **MUSCLES**

Muscles are composed of about 75% water and 20% proteins, with the remainder made up of stored carbohydrates and lipids for energy, inorganic salts, and non-protein compounds.

Skeletal muscle tissue
Skeletal muscles allow the body to move. They are voluntary muscles, controlled by the brain and spinal cord, and appear striped (striated) under a microscope.

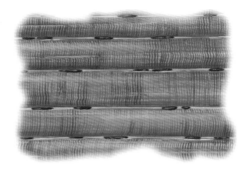

Smooth muscle tissue
Smooth muscle is controlled by the autonomic nervous system and is found in the skin, the blood vessels, and the reproductive and digestive systems.

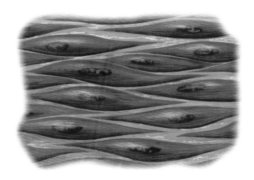

Cardiac muscle tissue
Cardiac muscle is the heart muscle, which contracts and relaxes rhythmically in an involuntary manner. It appears striped under a microscope.

Muscle fiber microstructure

SKELETAL MUSCLE TISSUE IS MADE UP OF MUSCLE FIBERS THAT ARE KNOWN AS MYOCYTES. Myocytes are composed of filaments of actin and myosin, which are responsible for contraction. Actin and myosin are arranged in repeating sections called sarcomeres. It is the interaction between the thin filaments of actin and the thicker filaments of myosin that produces movement.

Contraction of the muscle starts with an action potential that travels along the axon terminal of the motor neuron and crosses the neuromuscular junction, stimulating voltage-sensitive protein channels on the surface of the muscle cell membrane (called the sarcolemma). Along the surface of the sarcolemma and the t-tubule, the action potential is propagated as sodium ion voltage-gated channels open, causing an influx of sodium ions into the muscle cells. This influx of sodium ions depolarizes the plasma membrane, causing the sarcoplasmic reticulum to release calcium ions into the muscle cell cytoplasm. The calcium ions bind to troponin, which moves tropomyosin aside, exposing the binding sites on the actin protein for the binding of the myosin protein.

The next step in the muscle contraction cycle is the binding and release between myosin and actin within the sarcomere. When myosin binds to actin—the active site that has been exposed by the influx of calcium ions—it forms a crossbridge. This causes adenosine triphosphate (ATP) to be hydrolyzed into adenosine diphosphate (ADP) and inorganic phosphate (Pi), providing the energy necessary to change the angle of the myosin filament. The myosin filament pulls the actin filament, causing the sarcomere to shorten and the muscle to contract. Calcium is returned to the sarcoplasmic reticulum, and the binding sites on the actin filament are again covered, preventing further crossbridge cycling.

▶ **FROM BRAIN TO MUSCLE**

The process linking the nerve impulse to muscle movement is called excitation–contraction coupling. It is the process of converting an electrical stimulus to mechanical energy.

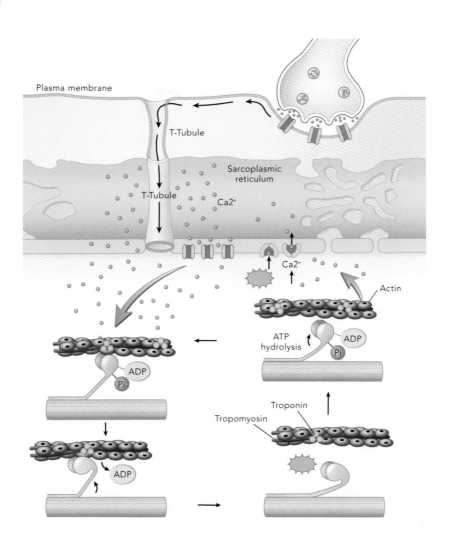

Plasma membrane

T-Tubule

T-Tubule

Sarcoplasmic reticulum

Ca2+

Ca2+

Actin

ATP hydrolysis

ADP

Pi

ADP

Pi

ADP

Troponin

Tropomyosin

Muscle types

THE SHAPE AND SITES OF
ATTACHMENT—THE ORIGIN (WHICH
DOESN'T MOVE DURING CONTRACTION)
AND THE INSERTION (WHICH DOES
MOVE DURING CONTRACTION)—
DETERMINE EACH MUSCLE'S FUNCTION.

The origin of a muscle is a specific site on a bone, usually one that is more proximal during a contraction. Generally, this bone has more mass than the bone at the insertion site. The insertion point is a specific site on a bone that is generally more distal. This is the bone that is most often moved by the contraction of the muscle. For example, when lifting an object, a contraction of the biceps brachii will move the insertion site on the radius closer to the origin sites on the scapula. However, in some situations the relationship is reversed, such as when performing a chin-up; the biceps brachii still contracts, but the origin is moved closer to the insertion that remains in place.

The action of a muscle is also determined by the direction in which the muscle fibers run. Fusiform muscles have fibers that predominantly run parallel to the length of the muscle. Unipennate muscles have fibers that attach to one side of a long tendon at an acute angle, thus resembling one half of a feather. Bipennate muscles consist of two rows of muscle fibers attaching at an acute angle on a central tendon, thus resembling a whole feather. Multipennate muscles have fibers arranged at multiple angles in relation to the direction of pull.

Pennation is the joining of muscles at an oblique angle to the tendon; the more oblique the angle, the greater the number of muscle fibers. This means that, given the same muscle size, a muscle that is bipennate, for example, is stronger than both unipennate muscle and fusiform muscle. While greater pennation increases the force that the muscle can produce, it also reduces the degree to which it can shorten, reducing range of motion.

There are a number of different specialized muscle shapes that reflect a specific function and method of attachment to the skeleton. For example, the pectoralis major attaches to the sternum at its origin and to the humerus at its insertion. The sternal attachment is spread out, while the insertion converges to a point, giving the muscle a fan or triangular shape. This also allows the muscle to pull the humerus in slightly different directions depending on which fibers are stimulated.

Circular muscles have a concentric arrangement of fibers, forming a ring shape. Contraction of this type of muscle makes the circle smaller and functions to close entrances to the body; an example of this is the orbicularis oris that is found in the lips.

▼ MUSCLE TYPES

(On the next pages) Illustrations of the various muscle types. Muscular contractions are controlled by the nervous system. Impulses from the nerves trigger the release of calcium in the cells of the muscle fibers, which initiates a contraction. The energy for muscular contraction comes from glycogen, a form of the sugar glucose, stored in muscle tissue.

Unipennate

Bipennate

Multipennate

Spiral

Spiral

Radial

Quadrilateral

Strap

Strap
(with tendinous intersections)

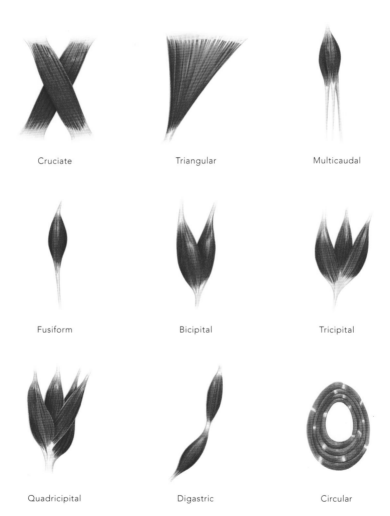

Cruciate

Triangular

Multicaudal

Fusiform

Bicipital

Tricipital

Quadricipital

Digastric

Circular

Muscle structure
Noncontractile components

THE CONTRACTILE COMPONENTS OF
MUSCLE ARE SURROUNDED AND
DIVIDED BY AN EXTENSIVE SET OF
EXTRACELLULAR CONNECTIVE TISSUES.
The connective tissues provide support
for the muscle and play a role in the
metabolic processes and transmission
of force to the tendon.

The epimysium is a tough layer
of connective tissue that is resistive
to stretching. It surrounds the entire
surface of the muscle, separating it
from other muscles. The perimysium
lies beneath the epimysium and divides
the muscle into bundles of fibers called
fascicles. Dividing the muscle this way
provides a conduit for blood vessels
and nerves. Like the epimysium, the
perimysium is tough and resistive to
stretching. The endomysium bundles
individual muscle fibers, lying just
outside the cell membrane. The
endomysium is where the blood vessels
exchange nutrients and waste products
with the muscle fibers. This layer of
connective tissue is more delicate, and
it connects with both the muscle fibers

and the perimysium. Though they are
described separately, in reality the
layers of connective tissue are
interwoven, forming part of a
continuous sheet of tissue and giving
strength, support, and elasticity to the
muscle. These tissues are continuous
with the tendon at the end of the
muscle, transferring contractile pulling
forces and providing tension when the
muscle isn't contracting, such as when
it is fully stretched.

▶ **NONCONTRACTILE TISSUE**
The elastic properties of
noncontractile tissue, shown here,
create a smooth translation of force
to the bone by dampening the effect
of the contraction. This helps to
stabilize the movement of the joint.

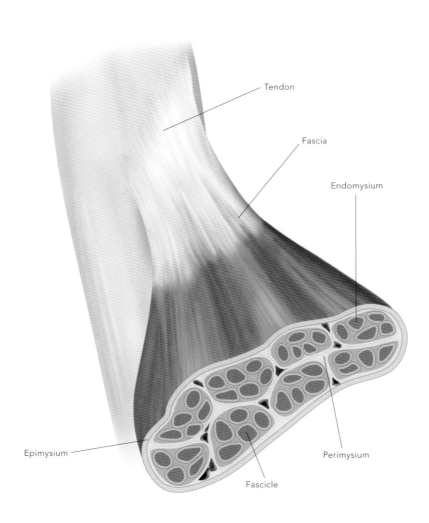

Tendon

Fascia

Endomysium

Epimysium

Perimysium

Fascicle

Contractile components

A WHOLE MUSCLE, SUCH AS THE DELTOID OR GASTROCNEMIUS, IS MADE UP OF MANY INDIVIDUAL FIBERS. Each fiber is actually an individual cylindrical cell with multiple nuclei. These are some of the largest cells in the body, with some up to 20 inches (50 cm) in length. Within each fiber are the building blocks—units called sarcomeres. A sarcomere, as described earlier, is made up of the actin and myosin filaments between two z-discs. Sarcomeres are aligned end to end along the length of the fiber, and when they shorten, they generate contraction of the whole fiber.

Structural proteins also play an important role in the transmission of force and the stabilization of the muscle structure. For example, titin is a very large protein that contributes to force transmission and helps to provide passive tension within the fiber when stretched. Another structural protein, desmin, stabilizes the alignment between the sarcomeres.

A single row of sarcomeres is called a myofibril. The sarcoplasmic reticulum wraps around and between each myofibril, where it regulates the flow of calcium during contraction. Myofibrils are bundled together into muscle fibers that are surrounded by the sarcolemma and endomysium. A number of fibers are bundled into compartments called a fascicle, surrounded by perimysium. These are then all surrounded by the epimysium to form the whole muscle.

▶ **MUSCLE FIBERS**

Observing the muscle fiber reveals that almost the entire cross section is taken up by myofibrils. Typically, there are hundreds of these packed into one muscle fiber.

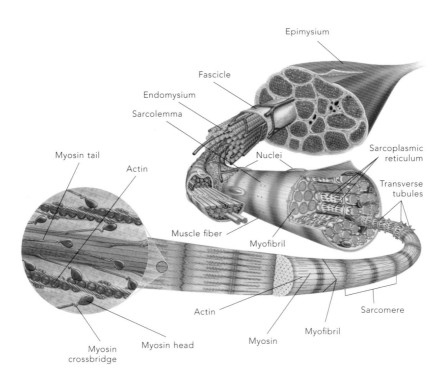

Epimysium

Fascicle

Endomysium

Sarcolemma

Nuclei

Sarcoplasmic reticulum

Transverse tubules

Myosin tail

Actin

Muscle fiber

Myofibril

Actin

Myosin

Myofibril

Sarcomere

Myosin crossbridge

Myosin head

Fiber types
Muscle fiber types

DEPENDING ON THE FUNCTION OF A SPECIFIC MUSCLE, THE DEMANDS PLACED ON THE MUSCLE THROUGH EXERCISE, AND THE GENETIC PREDISPOSITION OF AN INDIVIDUAL, A MUSCLE MAY HAVE A DIFFERENT MIX OF FIBER TYPES. There are a number of different types of muscle fibers; the three main types are referred to as Type I, Type IIa, and Type IIb. Structurally, these fibers are quite different in their composition.

Type I fibers are slow-contracting. These fibers are typically the smallest fiber type, but they have a high concentration of mitochondria, the cellular organelle responsible for energy production. They appear red in color due to the high density of capillaries to deliver blood and oxygen to the fiber and myoglobin to carry oxygen within the muscle.

Type IIa fibers are lighter in color than Type I fibers but still have a relatively high concentration of mitochondria. They have a high density of capillaries but are larger in size than Type I fibers.

Type IIb fibers are the largest of the three fiber types. Since they have a low capillary density and low levels of myoglobin, they are white in color.

▶ **TYPES OF MUSCLE FIBERS**
Type I, Type IIa, and Type IIb fibers are bundled adjacent to each other within the same fascicle, however, they may not be used in the same muscle contraction.

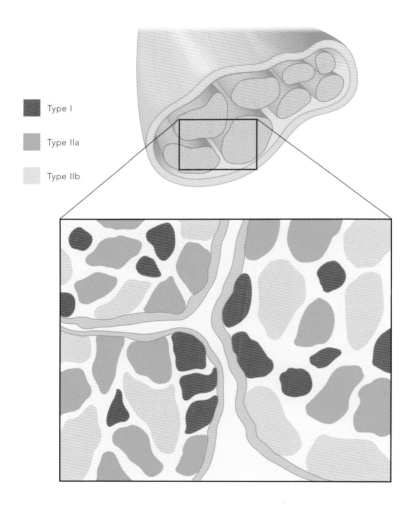

Type I

Type IIa

Type IIb

Muscle fiber properties

IF A MUSCLE IS TYPICALLY USED REPEATEDLY THROUGHOUT THE DAY, IT WILL LIKELY HAVE A HIGH PROPORTION OF TYPE I FIBERS. These fibers produce low levels of force and are slow to contract, but they do not fatigue easily. The capillary density and high number of mitochondria mean that these muscles are able to use oxygen when breaking down glucose and fat to make fuel for the muscle. These fibers are found in high proportions in leg muscles and trunk muscles that support posture. The types of tasks best suited for Type I fibers are activities such as a long walk or lifting a light object.

Type IIa fibers are called upon when more speed and strength is required, but only for a short period of time. These fibers produce medium to high levels of force and are fast to contract, but they fatigue more easily than Type I fibers. While still having a good blood supply for aerobic metabolism, these muscles also use anaerobic metabolic pathways.

The types of tasks best suited for Type IIa fibers are activities such as running faster or lifting a heavy object, as the long-term capacity of these fibers is small.

Type IIb fibers are called upon when maximal speed and strength is required, but only for a short period of time. These fibers produce high levels of force and are very fast to contract. However, Type IIb fibers fatigue quickly, as the main metabolic pathway is anaerobic. The types of tasks best suited for Type IIb fibers are activities such as a sprint or lifting a very heavy object.

▶ **FIBER COMPOSITION**

While genetics plays a role in whether a person will have a higher proportion of Type I (indicated opposite by the darker forms) or Type II fibers (indicated opposite by the lighter forms), exercise and occupational demands may cause the muscle to adapt. Those who are mostly sedentary will have a higher proportion of Type IIb muscle fibers.

Long-distance running

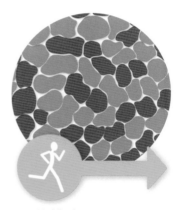

Middle-distance running

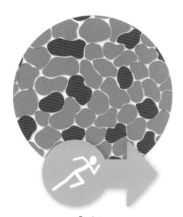

Sprint

Muscle contraction

Concentric contractions

MUSCLE CELLS ARE HIGHLY SPECIALIZED
FOR A SINGLE TASK: CONTRACTION.
There are three main types of muscle
contraction: concentric, eccentric,
and isometric.

In a concentric muscle contraction,
the tension in the muscle increases
until the muscle shortens. For example,
in order to lift an object in the left
hand, the muscle to contract is the
biceps brachii muscle of the left arm,
concentrically. Sarcomeres along the
length of the muscle shorten, and the
muscle itself contracts and shortens.
This flexes the elbow, generating force
that moves the lower arm upward.

▶ **MUSCLE RECRUITMENT**

During a concentric contraction, the
sarcomeres shorten. The first of the
lower two images shows a lengthened
sarcomere at the beginning of the
contraction; the second shows it
shortened at the end of the contraction,
after crossbridge cycling has occurred
to bring the ends closer together.
This is stimulated by the release of
calcium ion, and ATP is broken down
in the process.

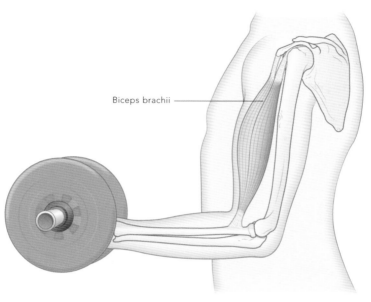

Biceps brachii

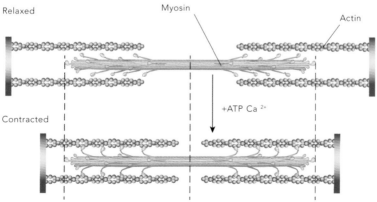

Relaxed

Myosin

Actin

+ATP Ca^{2+}

Contracted

Eccentric and isometric contractions

OTHER WAYS IN WHICH THE MUSCLE CAN CONTRACT AND GENERATE FORCE ARE KNOWN AS ECCENTRIC AND ISOMETRIC CONTRACTIONS. In an eccentric contraction, the muscle will generate force, but it is lengthening under tension rather than shortening. To lower an object previously lifted with the right arm, the biceps brachii muscle of the right arm has to contract eccentrically. Sarcomeres actually lengthen, thus allowing the elbow to extend and the lower arm that is holding the object to move downward. Crossbridge cycling of the myosin filament is still occurring and still generating force on the actin filament, but the force generated in the muscle is less than the force of the weight being lifted. Without the eccentric contraction to control the movement, the arms and object would just fall with gravity. We also use our quadriceps muscles eccentrically when running or walking downhill, to slow our descent.

In an isometric muscle contraction, the muscle maintains the same length even though tension in the muscle increases. Thus, there is no change in the length of the contracting muscle during this type of muscle contraction. If we want to hold our object stationary after lifting it, we have to isometrically contract the biceps brachii muscle of the right arm. Again, crossbridge cycling of the myosin filament is still occurring and still generating force on the actin filament, but the force generated by the muscle matches the force of the weight, and the elbow remains at the same height.

▶ **ECCENTRIC CONTRACTION**

During an eccentric contraction, the muscle is forcibly lengthened under load. When the load is high, damage to the muscle is incurred, resulting in inflammation and muscle pain. The inflammation peaks 24–48 hours after training. It is common after exercise involving eccentric contractions for the muscle to become stiff and sore.

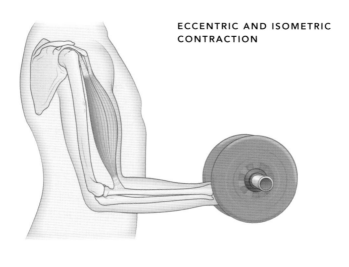

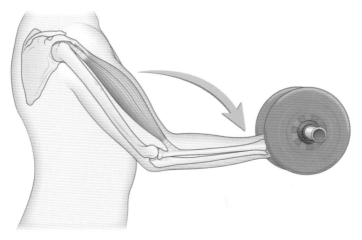

Length–tension relationship

SARCOMERES ARE THE SITE OF FORCE PRODUCTION GENERATED BY THE MUSCLE. That force is created by the head of the myosin filament binding to the actin filament and changing shape so that it produces a pulling motion. The amount of force produced during a contraction depends on a number of factors, including the number of sarcomeres running in parallel along the muscle, the number of fibers stimulated by the nerve, the pennation angle, and the length of the sarcomere or muscle.

The effect of sarcomere length on force production can be seen in the length–tension relationship curve. With the muscle at a resting length, the force produced by the sarcomere is highest. This is because there is a maximum number of myosin heads available to bind with the actin filament. There is no one optimal length; the muscle has a narrow range of optimal lengths over which tension does not vary with length and the active force generated is maximal.

If the muscle is lengthened beyond the optimal length, such as during a stretch, the force generated by the muscle will drop. This is due to a reduced number of myosin filaments overlapping with the actin filaments. In these areas, no crossbridges can form, and with greater lengthening, the degree of overlap and force is further reduced.

The force also declines when the sarcomeres are shorter than their optimal length. At moderately short lengths, the actin filaments begin to meet in the middle of the sarcomere and slide over each other, impairing the binding of myosin. Further shortening results in the myosin filament and the z-discs colliding, preventing any pulling action of the myosin head.

▶ **MUSCLE TONE**

Maximal tension is readily produced as the body maintains resting muscle length near the optimum part of the length–tension curve. It does this by maintaining a muscle tone, keeping the muscles partially contracted.

LENGTH–TENSION RELATIONSHIP

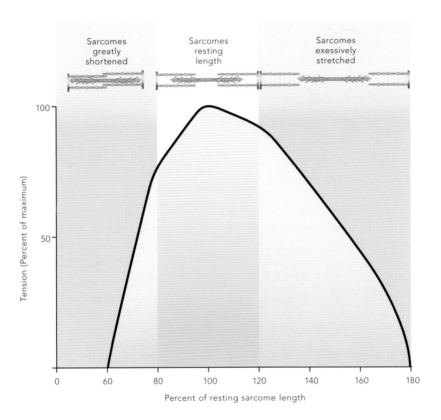

Training and detraining
The effects of training

MOST TISSUES IN THE BODY ARE ABLE TO ADAPT TO THE DEMANDS PLACED ON THEM. For example, if bones are repeatedly placed under heavy loads, they will become denser. In addition, the hands develop calluses if they are subjected to a lot of manual work. Muscles are able to adapt in response to stresses as well, with different demands driving different adaptations. The stresses that cause them may be due to occupational, environmental, or exercise demands.

Resistance training is a specific form of exercise that involves the successive lifting of weights. It is used for many purposes, including improving athletic performance, aiding recovery from injury, and increasing independence and function among older adults. The main adaptations involved are an increase in the cross-sectional area of the whole muscle and of individual muscle fibers. This is due to an increase in myofibril size and number and is referred to as hypertrophy. Satellite cells are a skeletal muscle precursor and are activated in the very early stages of training, fusing with existing fibers to increase their size. Other adaptations include a change in fiber type toward an increase in Type IIa fibers, increased myofilament density, and strengthening connective tissue and tendons.

Aerobic exercise such as walking, running, and cycling elicits different adaptations in the muscle. These adaptations are primarily to make the muscle more efficient at providing energy during sustained work. One of the main biochemical adaptations induced by aerobic training is an increase in the number of mitochondria throughout the trained muscle fibers. This increases the capacity for aerobic energy provision from fat and glucose. This adaptation is primarily seen in the Type I fibers. Aerobic exercise also increases the number of capillaries surrounding individual muscle fibers, increasing the flow of blood and oxygen to the muscle.

RESISTANCE TRAINING

BEFORE   **AFTER**

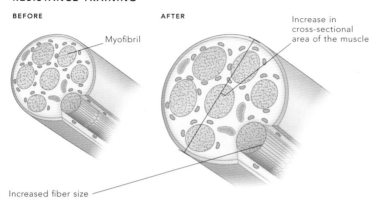

Myofibril

Increase in cross-sectional area of the muscle

Increased fiber size

Increased number of mitochondria and blood vessels

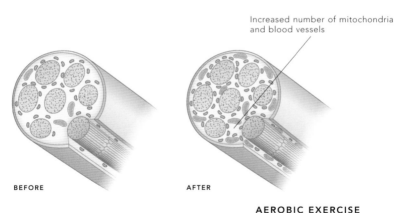

BEFORE   **AFTER**

AEROBIC EXERCISE

▲ **HYPERTROPHY**

An increase in myofibril size is called hypertrophy. With weight training, all fiber types will increase in size.

The effects of detraining

THE ADAPTATIONS WE SEE AS A RESULT OF EXERCISE WILL ONLY BE MAINTAINED IF THE MUSCLE IS CONTINUALLY EXPOSED TO A SUFFICIENT STIMULUS. As a result, if exercising is stopped, these adaptations will disappear—this is referred to as detraining.

Most individuals typically lose a minimum of 8 percent of muscle mass every decade after the age of 40. This muscle loss is not inevitable and can be reversed or reduced with appropriate exercise. With resistance training, muscles may increase in cross-sectional area by up to 15 percent after three months of training. However, muscle size will return to pre-training levels within a similar period of time if the stimulus is removed. The muscle will also get stronger as a result of training. Strength gains occur rapidly in those who start exercising for the first time, with around 30 percent improvement seen over the first three months.

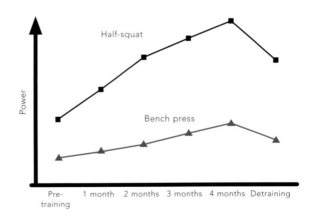

Increases in strength are not lost as rapidly as muscle size, but muscle power—the ability to generate force—is lost quickly through detraining. Adaptations in the muscle, brought about by aerobic exercise, are also lost through detraining. A number of factors, including changes to the heart, occur that will result in a loss of ability to exercise at the same level.

Atrophied muscle

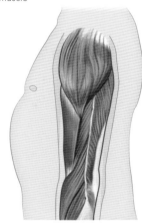

Hypertrophied muscle

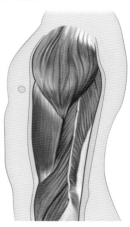

◀ **STRENGTHENING EXERCISES**

A rapid increase in strength is noticeable in those taking up exercise for the first time. Significant gains can be achieved in as little as three months.

▶ **MUSCLE QUALITY**

Muscle quality refers to the force a muscle produces relative to its size. Aging and chronic disease lead to a decrease in muscle quality, while strength training can improve both muscle size and quality.

Force and velocity

DURING CONTRACTION THERE IS A TRADE-OFF BETWEEN THE AMOUNT OF FORCE GENERATED AND THE SPEED OF MOVEMENT. While under a light load, a muscle can contract and shorten very rapidly. However, as the load gets heavier, the velocity at which the muscle can shorten is reduced. As with the length–tension relationship, the reduction in force is caused by the reduced opportunity for crossbridging between the myosin and the actin. It takes a certain amount of time for crossbridges to form and for filaments to slide past one another at a faster rate. Therefore the force decreases due to the lower number of crossbridges that are able to form.

The slower the rate of muscle shortening, the stronger the muscle contraction is. Thus, an isometric contraction produces more force than a concentric contraction. This is because the muscle is not lengthening during an isometric contraction. Likewise, the greatest amount of force exerted is during an eccentric contraction in most cases. For example, it is possible to lower a heavy weight under control—for example, from a bench down to the floor—but it would not be able to be lifted back up again. In eccentric contractions, each myosin crossbridge exerts more force as it is pulled apart and detached. As the crossbridge reattaches more rapidly, the result is the production of a higher force.

This relationship between force and velocity affects the rate at which muscles can perform mechanical work (power). Since power is equal to force times velocity, the muscle generates little to no power with either an isometric contraction (due to zero velocity) or at maximal velocity (due to negligible force).

▶ **TRAINING FOR POWER**

As speed of movement is so important in many sports, athletes and coaches will adapt training programs to cause a shift in the force–velocity curve to the right—being able to exert force faster.

FORCE AND VELOCITY DURING MUSCLE CONTRACTION

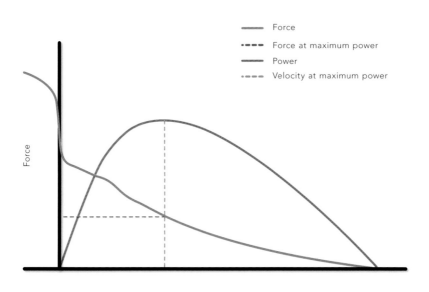

Force

— Force
---- Force at maximum power
— Power
---- Velocity at maximum power

Velocity

SECTION TWO:

NEUROLOGICAL BASIS FOR MOVEMENT

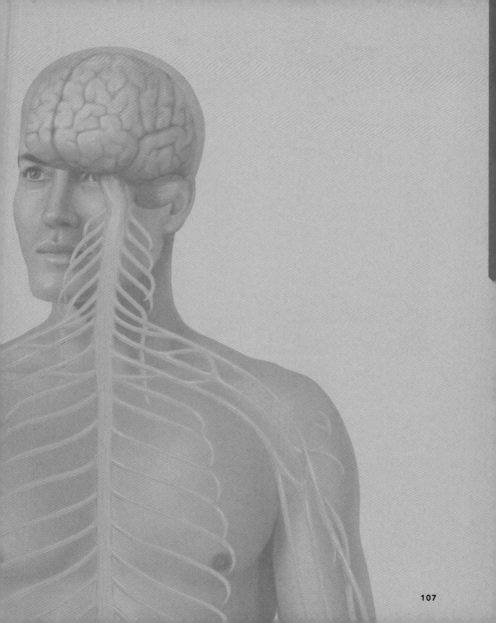

Chapter 5:
Structure and function of the nervous system

In this chapter, we explore the role of nerves as the body's message carriers, transmitting signals from the brain that stimulate movements or reactions, while also carrying messages back to the brain that may trigger a conscious or subconscious response. Nerves originate from the brain, and run through the spinal cord to peripheral parts of the body, controlling the movements that are an integral part of life.

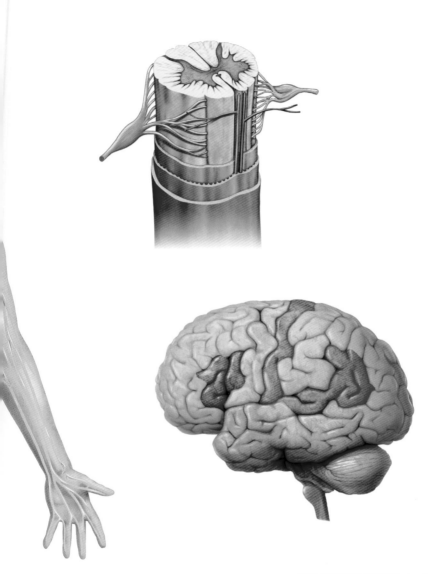

Nervous system overview

THERE ARE THREE BIOLOGICAL SYSTEMS
WITHIN THE HUMAN BODY THAT ARE
RESPONSIBLE FOR ITS HEALTH AND
FUNCTION: THE ENDOCRINE, IMMUNE,
AND NERVOUS SYSTEMS.

The nervous system influences, and is influenced by, its companion systems. It has evolved to identify changes in the external (conditions outside the body) and internal (conditions within the body) environments. The nervous system then decides how to respond with quick and appropriate signaling to its muscles, organs, and glands.

From detecting changes in internal body temperature to reacting to external stimuli, the nervous system is structured to identify, process, react, and store memories. Like most systems in the body, the nervous system needs to be challenged physically and mentally to help maintain its structure and function.

▶ **COMPONENTS OF THE NERVOUS SYSTEM**

The nervous system comprises the central nervous system (CNS) and peripheral nervous system (PNS). These two systems form a complex network that allows the body to detect and respond to internal and external changes.

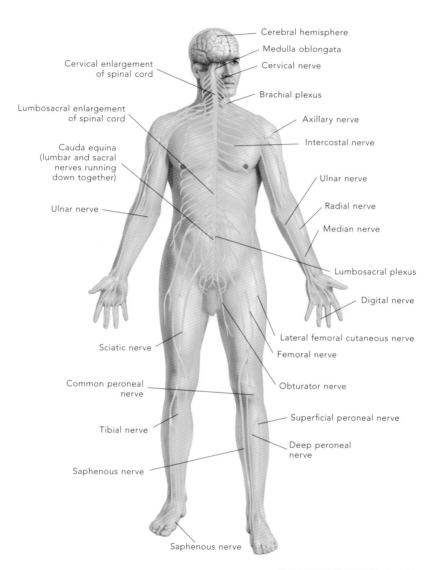

Cerebral hemisphere

Medulla oblongata

Cervical nerve

Cervical enlargement of spinal cord

Brachial plexus

Lumbosacral enlargement of spinal cord

Axillary nerve

Intercostal nerve

Cauda equina (lumbar and sacral nerves running down together)

Ulnar nerve

Radial nerve

Median nerve

Ulnar nerve

Lumbosacral plexus

Digital nerve

Sciatic nerve

Lateral femoral cutaneous nerve

Femoral nerve

Common peroneal nerve

Obturator nerve

Superficial peroneal nerve

Tibial nerve

Deep peroneal nerve

Saphenous nerve

Saphenous nerve

Components of the nervous system

THE NERVOUS SYSTEM IS COMPOSED OF THE CENTRAL NERVOUS SYSTEM (CNS)—the brain and spinal cord—and the peripheral nervous system (PNS), which connects the CNS with the body. The PNS is functionally divided into the somatic nervous system (SNS) and the autonomic nervous system (ANS).

The CNS is often regarded as the control center of the nervous system, as this is where most information is received and distributed. The CNS consists of the brain and spinal column, each housed within their own protective case, the cranium and vertebral canal (spinal column), respectively.

Both of these structures are encased in three membranous coverings known as the meninges. Like the cranium and vertebral canal, the role of the meninges is to protect the CNS. The most superficial membrane is the dura mater, followed by the arachnoid mater and then the pia mater—the deepest of the three. The outermost membranes, the dura and the arachnoid mater, surround the brain loosely when compared with the pia mater, which is thinner and firmly adhered to the brain.

The PNS is the link between the CNS and the rest of the body. It communicates data from the internal organs and external environment to the CNS and then delivers "orders" from the CNS to the muscles, organs, or the glands.

▶ **THE CENTRAL NERVOUS SYSTEM**

Right: The brain and spinal column; far right: a cross-section. The brain and spinal column make up the central nervous system. They are protected by membranous coverings.

THE CENTRAL NERVOUS SYSTEM

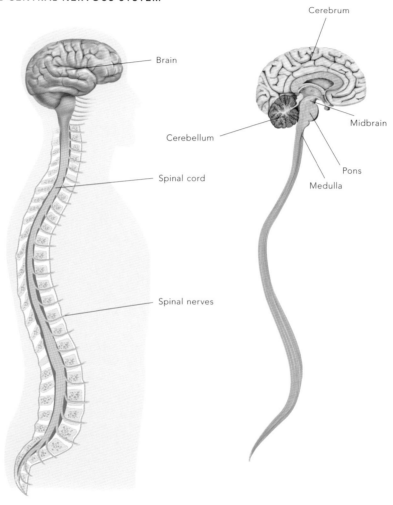

Brain

Cerebrum

Spinal cord

Cerebellum

Midbrain

Pons

Medulla

Spinal nerves

Autonomic nervous system

THE AUTONOMIC, OR INVOLUNTARY, NERVOUS SYSTEM, IS RESPONSIBLE FOR MAINTAINING THE BODY'S INTERNAL FUNCTIONAL STATE (HOMEOSTASIS). It innervates smooth muscle, cardiac muscle, and glands, making necessary changes to maintain homeostasis. Changes include increasing blood flow to an organ and slowing the heartbeat, among many other internal processes.

There are two divisions that effect these involuntary changes, the sympathetic and parasympathetic divisions. These competing divisions act upon the same organs to either increase or decrease excitability. The sympathetic division is responsible for readying the body for activity, increasing the heart and respiratory rate, and exciting the organs that can help with the fight or flight reflex. The fibers of the sympathetic division originate from the thoracolumbar region of the spine.

In contrast, the parasympathetic division helps the body relax, conserve energy, and maintain homeostasis.

Unlike the sympathetic division, the parasympathetic's fibers originate from the craniosacral region of the spine.

▶ **COMPOSITION OF THE AUTONOMIC NERVOUS SYSTEM**

The motor portion of the ANS helps to regulate the activity of smooth muscle, cardiac muscle, and glands. The ANS consists of the sympathetic and parasympathetic divisions. Both divisions usually have opposing effects on organs. For example, the sympathetic division will increase (excite) heart and respiratory rate, whereas the parasympathetic division will decrease (inhibit) both.

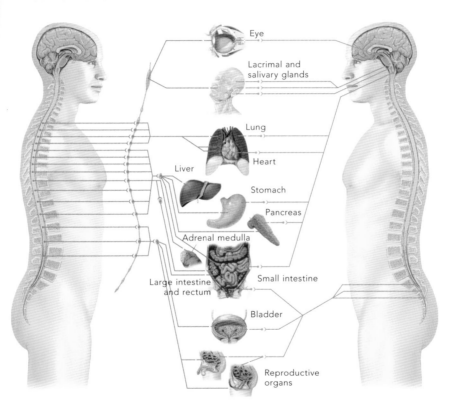

Eye

Lacrimal and
salivary glands

Lung

Heart

Liver

Stomach

Pancreas

Adrenal medulla

Small intestine

Large intestine
and rectum

Bladder

Reproductive
organs

Brain overview

THE HIGHER FUNCTION OF THE HUMAN BRAIN IS WHAT SEPARATES US FROM OTHER SPECIES. The human brain has the ability to receive sensory information and interpret and integrate this information to provide a motor output. It keeps us alive by controlling all of our vital functions, from breathing to smooth muscle contraction in the gut to aid digestion, and provides functions such as thought, memory, and language.

The brain is structurally complex and can be divided into three distinct areas: the cerebrum, the cerebellum, and the brainstem. The brainstem is the most central structure of the brain and consists of the medulla, pons, and midbrain. The cerebellum, the largest part of the hindbrain, covers this stalk-like part of the brain posteriorly. It connects to the brainstem via three pairs of fiber bundles, the inferior, middle, and superior peduncles, all connecting to the three regions of the brainstem. The cerebrum, the largest structure of the forebrain, extends posteriorly over the cerebellum. This is the principal part of the brain that contains two identical hemispheres, the left and right. The most superficial layer of the cerebral hemisphere, the cerebral cortex, is constructed of gray matter. The gray matter of the cortex contains a large number of nerve cell bodies.

▶ **OVERVIEW OF THE BRAIN**
The brain is where information is received and then distributed. It is the control center of the nervous system, where billions of neurons interact to keep our body functioning.

Gyri
These are the ridges formed by the folded surface of the cerebral cortex.

Sulci
These are grooves or furrows in the folded surface of the cerebral cortex.

Cerebrum
With an inner core of white matter and an outer cortex of gray matter, the cerebrum is the largest part of the brain.

Brainstem
Consisting of three parts, the midbrain, pons, and medulla, the brainstem is continuous with the spinal cord below. The brainstem is involved in regulation of such vital functions as breathing, heartbeat, and blood pressure.

Cerebellum
Like the cerebrum, the cerebellum has a highly folded outer layer. The cerebellum plays an important role in controlling movement, coordinating voluntary muscular activity, and maintaining balance and equilibrium. Lying at the base of the cerebrum, it is attached to the brainstem.

Lobes of the brain

THE LONGITUDINAL FISSURE SEPARATES THE TWO HEMISPHERES OF THE CEREBRUM. Each hemisphere can then be further divided into four lobes, the borders of which are marked by folds (sulci) and smooth areas (gyri). The four lobes of the brain are the frontal lobe, parietal lobe, temporal lobe, and occipital lobe. The frontal lobe is the largest, while the parietal and temporal lobes are situated directly posteriorly and caudally, respectively.

The most posteriorly located lobe is the occipital lobe, which is separated from the parietal lobe by the parieto-occipital sulcus.

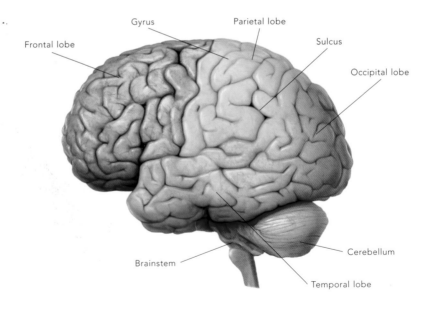

Gyrus

Parietal lobe

Frontal lobe

Sulcus

Occipital lobe

Cerebellum

Brainstem

Temporal lobe

Functional areas of the brain

THE OUTER LAYER OF THE CEREBRUM IS CALLED THE CEREBRAL CORTEX AND PLAYS A CRITICAL ROLE IN HIGHER FUNCTIONS, such as consciousness, thought, language, and memory. Key functional areas of the cerebral cortex include the primary motor cortex (frontal lobe), the somatosensory cortex (parietal lobe), the auditory cortex (temporal lobe), and the visual cortex (occipital lobe).

The functional areas of the cerebral cortex are often referred to by their numbered divisions, which were assigned by Korbinian Brodmann in 1909. Brodmann discovered these functional areas by analyzing the arrangement of cells, or cytoarchitecture. Since then, scientists have concluded that these areas have diverse functions that connect and interact with each other.

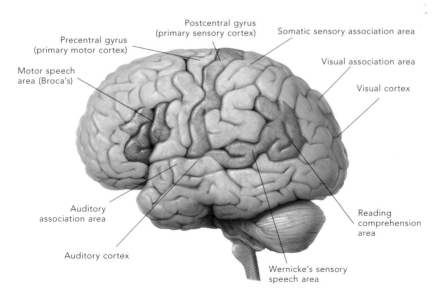

Postcentral gyrus (primary sensory cortex)

Precentral gyrus (primary motor cortex)

Somatic sensory association area

Visual association area

Visual cortex

Motor speech area (Broca's)

Auditory association area

Auditory cortex

Reading comprehension area

Wernicke's sensory speech area

The spinal cord

THE SPINAL CORD IS AN EXTENSION OF THE BRAINSTEM THAT ORIGINATES AT THE FORAMEN MAGNUM (base of the skull) and extends inferiorly down through the vertebral canal until it terminates into the conus medullaris (L1 and L2 vertebrae in adults).

When transversely dissected, the spinal cord reveals a central and an outer segment. The central segment is termed gray matter. Gray matter is the grouping of neurons and glial cells. The H-shaped gray matter has four horns, two projecting posteriorly and two projecting anteriorly, named the dorsal and ventral horns, respectively. The dorsal horn contains groups of neurons associated with sensory functions. The ventral horns give rise to motor neuron axons that innervate the skeletal muscle; therefore, they are often termed the motor portion of the gray matter.

The outer segment of the spinal cord is called white matter. White matter has three regions called funiculi, each named according to its location: posterior, lateral, and anterior. The posterior funiculus contains myelinated nerve fibers associated with positional sense, movement, and touch sensation. The lateral and anterior funiculi contain tracts concerned with proprioception (posterior spinocerebellar), transmission of pain and temperature (lateral spinothalamic), and light touch (anterior spinothalamic).

There are 31 pairs of spinal nerves, each named and numbered according to its emergence from the vertebral canal: eight cervical, twelve thoracic, five lumbar, five sacral, and one coccygeal. The first seven spinal nerves exit above their corresponding vertebra; since the first spinal nerve leaves above the first cervical vertebra (the atlas), there are eight nerves for seven vertebrae. The remainder of the spinal nerves exit below their corresponding vertebra. At the conus medullaris, the remainder of the spinal nerves continue inferiorly down through the spinal column in the form of the cauda equina (Latin for "horse's tail").

SPINAL CORD WITHIN THE MENINGES

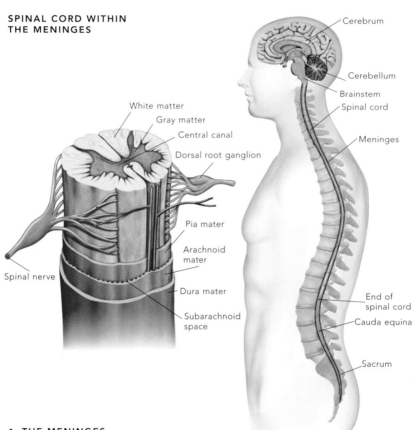

White matter

Gray matter

Central canal

Dorsal root ganglion

Pia mater

Arachnoid mater

Spinal nerve

Dura mater

Subarachnoid space

Cerebrum

Cerebellum

Brainstem

Spinal cord

Meninges

End of spinal cord

Cauda equina

Sacrum

▲ THE MENINGES

The spinal cord extends inferiorly from the brain and is encased within three membranous coverings known as the meninges.

Spinal nerves

ALL SPINAL NERVES—EXCEPT THE FIRST PAIR—EMERGE FROM THE SPINAL COLUMN THROUGH OPENINGS IN THE VERTEBRAE KNOWN AS INTERVERTEBRAL FORAMINA. There are 31 spinal nerves on the left and right sides of the vertebrae, and each nerve has two roots. These paired roots are the dorsal and ventral roots, which carry afferent sensory fibers and efferent motor fibers, respectively. Each spinal nerve connects to the spinal cord through small bundles of nerve fibers called rootlets. Functionally, a spinal nerve is formed when the dorsal and ventral roots meet, producing a mixed nerve.

When the spinal nerve leaves the intervertebral foramen, it will separate into several branches called rami. The thicker ventral ramus passes anteriorly to innervate the skin and muscle on the ventral aspect of the trunk, neck, and extremities. In comparison, the smaller dorsal ramus passes posteriorly to innervate the muscle and skin on the back.

Unlike spinal nerve roots, the ventral and dorsal rami possess both sensory and motor fibers. Ventral rami combine at the cervical and lumbosacral regions to form networks called neural plexuses. Individual nerves arising from different neural plexuses are attributed to innervation of the head, neck, and upper and lower limbs. There are five neural plexuses in the human body, two for the upper body and three for the lower body. The cervical plexus (C1–C4) innervates the head, neck, and shoulders. The brachial plexus (C5–T1) innervates the chest, shoulders, and upper limbs. The lumbar plexus (L1–L4) and the sacral plexus (L4–S4) are collectively known as the lumbosacral plexus and innervate the lower back, abdomen, groin, buttocks, and lower limbs. The coccygeal plexus (S4–Co1) innervates the small region over the coccyx.

▶ **CLASSIFICATION OF SPINAL NERVES**

Spinal nerves are named according to the region of the spine where they emerge.

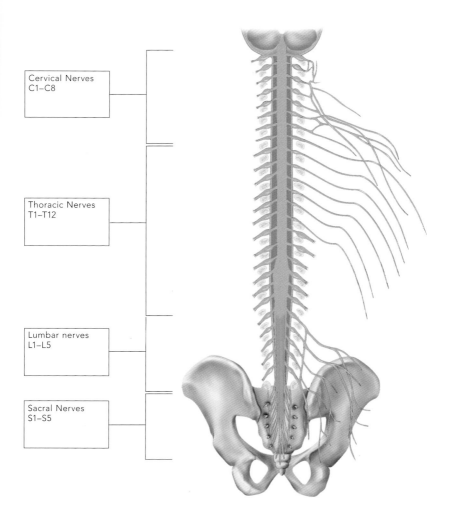

Cervical Nerves
C1–C8

Thoracic Nerves
T1–T12

Lumbar nerves
L1–L5

Sacral Nerves
S1–S5

Dermatomes and myotomes

SPINAL NERVES ARE COMPOSED OF
SENSORY AND MOTOR FIBERS THAT
INNERVATE SKIN AND MUSCLES. The
body can be divided into areas or
regions depending on what set of
nerve fibers innervate that region.
An area of skin innervated by a single
spinal nerve is known as a dermatome,
and a group of muscles that are
innervated by a single spinal nerve is
known as a myotome.

Dermatomes are strictly defined
bands around the body and are
specified according to the spinal
nerve that innervates each one.
Understanding the dermatomal
patterns is important when assessing
for injury to the nerve root, as
sensation can be partially or fully
disrupted. An example of a dermatome
is the skin of the back of the head
and around the neck, which is
supplied by C2 and C3—two spinal
nerves of the cervical plexus. The rest
of the cervical spinal nerves (C4–C8)
supply the shoulders, arms, and hands,
whereas the thoracic spinal nerves

from T2–T12 innervate the
interthoracic space.

Myotomes are groups of muscles
that are innervated primarily by a
specific spinal nerve. However, spinal
nerves can have a segmental origin
whereby multiple spinal nerve roots
supply nerve fibers to a single spinal
nerve. An example of this is the
musculocutaneous nerve—which
innervates the biceps brachii—that has
nerve roots at C5, C6, and C7 within
the brachial plexus. In contrast,
laterally or medially abducting the
fingers of the hand requires just a
single nerve root at T1 within the
brachial plexus.

▶ **DERMATOMES**
Each spinal nerve innervates an area of
skin known as a dermatome.

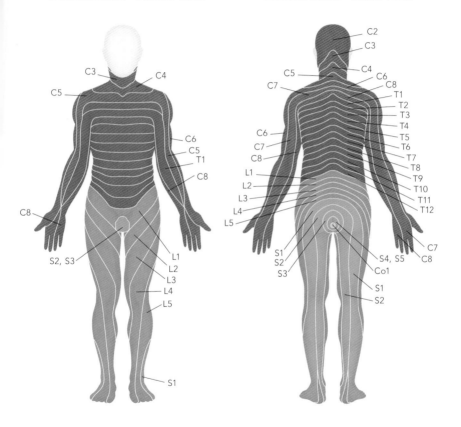

DERMATOMES—FRONT VIEW

DERMATOMES—REAR VIEW

C3
C4
C5
C6
C5
T1
C8
C8
S2, S3
L1
L2
L3
L4
L5
S1

C2
C3
C4
C5
C6
C7
C8
T1
T2
T3
T4
T5
T6
T7
T8
T9
T10
T11
T12
C6
C7
C8
L1
L2
L3
L4
L5
S1
S2
S3
C7
C8
S4, S5
Co1
S1
S2

Overview of nerve cells

A NEURON, OR NERVE CELL, IS A SPECIALIZED CELL DESIGNED TO RECEIVE AND TRANSMIT INFORMATION WITHIN THE NERVOUS SYSTEM. These functional units are unique to the nervous system and are designed to transmit impulses over large distances (sometimes up to 3 feet/1 m) without compromising the strength of the information.

The cell body of a neuron (soma) contains the cell nucleus and the majority of cellular organelles. Key features of the neuronal soma are the presence of Nissl granules—cellular bodies made up of ribosomes and rough endoplasmic reticulum—which are the sites of protein synthesis. There are typically two processes that project outward from the cell body: dendrites and axons. Dendrites are short, with large surface areas that receive impulses and convey them to the cell body. Dendrites have small protrusions called spines, which help to increase the surface area of the dendrite. The axon, however, is responsible for transporting impulses away from the cell body to surrounding neurons or muscle. Axons vary in length and can be insulated by a layer of myelin. Myelin is a multi-layered sheath of phospholipids that gives white matter its color. Myelin is produced by Schwann cells in the peripheral nervous system and by oligodendrocytes in the central nervous system. Myelination insulates the axons and increases the speed of electrical impulses along an axon.

▶ **UNI- AND MULTIPOLAR NEURONS**

Multipolar neurons are the most common; they include interneurons and motor neurons. Sensory neurons are classified as unipolar neurons.

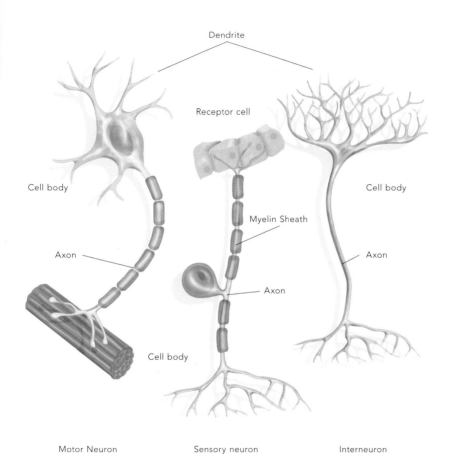

Dendrite

Receptor cell

Cell body

Myelin Sheath

Cell body

Axon

Axon

Cell body

Axon

Cell body

Motor Neuron

Sensory neuron

Interneuron

Neurotransmitters

NEURONS COMMUNICATE WITH EACH
OTHER THROUGH THE RELEASE OF
MESSENGER MOLECULES KNOWN AS
NEUROTRANSMITTERS.

Neurotransmitters are chemical
signals that are transmitted across a
synapse and are only released when an
action potential (see page 130) is
reached in the neuron. There are two
different types of neurotransmitters:
small-molecule neurotransmitters and
peptide neurotransmitters. Both of
these groups interact with ionotropic
and metabotropic receptors.

Small-molecule neurotransmitters
include amino acids such as glutamate
and glycine and are part of a precise
system of presynaptic release and
postsynaptic receptor activation. These
systems form the majority of excitatory
signaling in the central nervous system
(CNS) and are important for fast and
precise impulses in the sensory and
motor tracts. Another important
neurotransmitter is acetylcholine,
which, like glutamine, is a small-
molecule neurotransmitter. However,
acetylcholine is a neurotransmitter
involved in the autonomic nervous
system (ANS) as well as in the brain.
In the ANS, it is the primary
neurotransmitter responsible for
muscular contraction, whereas within
the CNS, it is involved with cognitive
roles such as attention.

▶ **NEUROTRANSMITTER PATHWAYS**
Neurotransmitters (in synaptic
vesicles) are transported to the end
of the axon and released into the
synaptic cleft between the axon
and the next nerve cell.

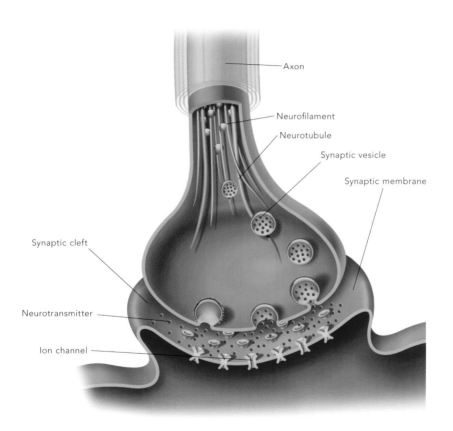

Axon

Neurofilament

Neurotubule

Synaptic vesicle

Synaptic membrane

Synaptic cleft

Neurotransmitter

Ion channel

Action potentials

AT REST, A NERVE CELL CONTAINS A NEGATIVE CHARGE OF AROUND 70 MILLIVOLTS; THIS IS KNOWN AS ITS RESTING POTENTIAL. When a nerve cell is stimulated, the cell reverses its polarity—depolarization—by opening voltage-gated ion channels within the cell membrane. The influx of positively charged sodium ions generates an electrical current across the membrane, which causes an electrical impulse—an action potential—to travel down the axon. The cell has to be stimulated above a certain level to create an action potential, and this is known as the threshold stimulus. After an action potential has occurred, there is a small period of time in which it is impossible for another action potential to be generated, and this is called the absolute refractory period (usually 0.5–1.0 millisecond).

The place where one nerve cell meets another is called a synapse. These specialized junctions can be either electrical or chemical, with varying speeds of transmission. Chemical synapses are characterized as slower and more abundant in our bodies compared with the faster electrical synapses located in the brain and retina. The cell sending the signal is known as the presynaptic nerve cell, and the cell receiving the signal is known as the postsynaptic nerve cell. These signals are sent across the gap between the cells—known as the synaptic cleft—to the postsynaptic nerve cell, which converts it back to an electrical signal (action potential).

▶ **RESTING AND ACTION POTENTIAL**
During its resting potential the nerve cell is negatively charged and its membrane channels are closed. During an action potential the membrane channels open, allowing for positively charged sodium to flow into the axon, increasing its charge.

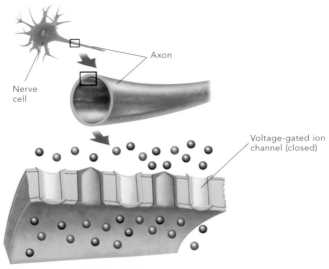

Axon

Nerve cell

Voltage-gated ion channel (closed)

INTERIOR OF AXON

Sodium ion channels closed
before or after action potential

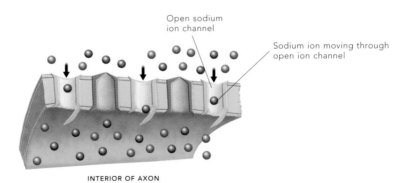

Open sodium
ion channel

Sodium ion moving through
open ion channel

INTERIOR OF AXON

Sodium ion channels open
during action potential

Sensory and motor pathways

NEURONS THAT RESIDE IN THE DORSAL ROOT GANGLION INNERVATE THE SENSORY STRUCTURES (SUCH AS THE SKIN) VIA THEIR AXONS AND ARE CLASSIFIED AS FIRST-ORDER NEURONS. The first-order neuron enters the central nervous system (CNS), where it will synapse with a second-order neuron. The second-order neuron will project further up the CNS to a third-order neuron. Finally, the third-order neuron will synapse with the complex neural network within one of the brain's lobes (e.g., the cerebral cortex). Due to the direction the impulse travels (receptor surface to the brain), this neuronal pathway is termed the ascending sensory tract. The spinocerebellar tracts are an example of ascending sensory tracts. These tracts supply the cerebellum with proprioceptive information from the muscles.

The motor component of the brain will integrate the sensory signals from the ascending tract. This motor network will then produce an impulse to a nerve fiber that will project to a neuron that supplies an effector (e.g., muscle). This neuronal pathway's impulse is descending in direction (brain to muscle); therefore it is known as the descending motor tract. The corticobulbar and corticospinal tracts are descending motor tracts that are responsible for sending impulses to the voluntary muscles of the body.

▸ **SENSORY AND MOTOR PATHWAYS**

This illustration shows the course of a sensory and motor pathway as it ascends and descends the nervous system.

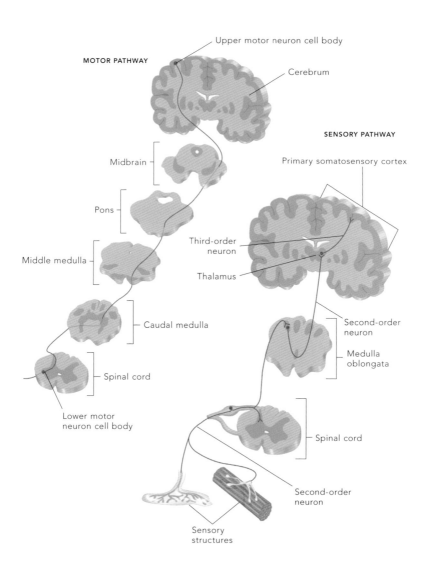

MOTOR PATHWAY

Upper motor neuron cell body

Cerebrum

SENSORY PATHWAY

Primary somatosensory cortex

Midbrain

Pons

Third-order neuron

Thalamus

Middle medulla

Caudal medulla

Second-order neuron

Medulla oblongata

Spinal cord

Lower motor neuron cell body

Spinal cord

Second-order neuron

Sensory structures

Spinal cord injury

Although rare, spinal cord injury is life-threatening and often leaves an individual with a loss of function to a specific area of the body.

The major difference between nerve cells in the central nervous system (CNS) and peripheral nervous system (PNS) is their ability to regenerate. Nerve cells in the CNS do not have the capacity to regenerate; therefore, once they have become injured or have died, they are unable to function.

Although the spinal cord is well protected, injury to its structure is possible as a result of a break or dislocation (severing the spinal cord), pressure, bruising, or a reduction in blood flow. The effect these injuries will have on function is dependent upon the level at which the injury occurred.

Function is lost below the level of the injury due to a failure in communication between the CNS and the periphery. The injury can be complete, when no fibers below the injury are functioning, or incomplete, when some but not all nerve fibers are functional.

Just over half of all spinal cord injuries result in loss of sensation and paralysis in both the upper and lower limbs, a condition known as tetraplegia (or quadriplegia). This occurs when an individual suffers an injury to the cervical spine.

If the injury occurs below the cervical spine, the upper limb will remain functional, but the lower limb will lose function, a condition known as paraplegia.

▶ **LOSS OF FUNCTION**

Injury to the spinal cord at the level of C4 will result in a loss of function of the upper and lower limbs (tetraplegia), whereas an injury at the level of T1 will only result in the loss of function of the lower limbs (paraplegia).

PARAPLEGIA

TETRAPLEGIA

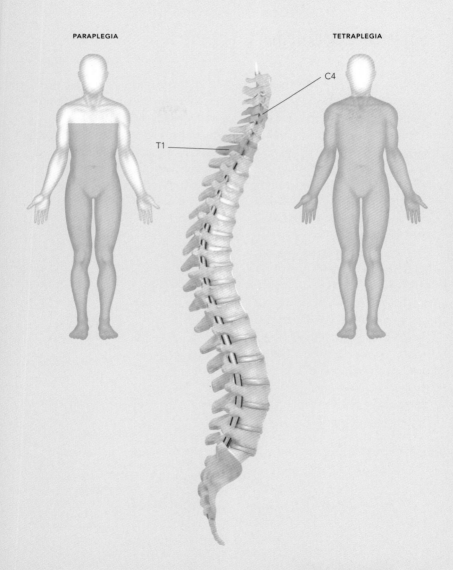

C4

T1

Spinal cord injury (*cont.*)

Modern advancements in medical science have greatly increased the survival rate of individuals with spinal cord injuries. However, the progression toward neural repair and regrowth has not been as successful. After spinal cord injury, there are several barriers to axon regrowth. One obstruction to axon regeneration is the formation of a glial scar. A glial scar impedes axon regrowth through the release of inhibitory molecules, as well as scar tissue being nonfunctional. Removing the glial cells completely is not viable, as removal of the cells produces widespread excitotoxicity as well as a decrease in important neurotrophic factors.

A lot more research is required to determine how to achieve an optimal environment for axon repair and regeneration. It is likely that successful treatments of spinal cord injury will consist of various therapies that promote regeneration.

▶ **AXON REGENERATION**

A demonstration of how nerve cells could one day regrow axons artificially through the introduction of support cells (shown here as red circles in phase 1).

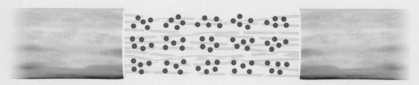

1. Fluid phase
Accumulation of neurotrophic factors
and extracellular matrix molecules

2. Cellular phase
Cell migration, proliferation, and alignment,
and axon formation

3. Axonal phase
Growth of axons

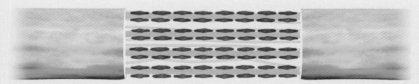

4. Myelination phase
Myelination of regenerated immature axons
forming mature axonal fibers

Chapter 6:
Sensory systems

This chapter shows how the senses help to fine-tune our movements, allowing the brain to determine the position of the limbs within the context of the external environment. It shows how the feedback received from the sensory system helps the body to produce highly skilled movements subconsciously and often at high speed. Receptors and nerves send signals to the brain, and their role in regaining controlled movement after injury is crucial.

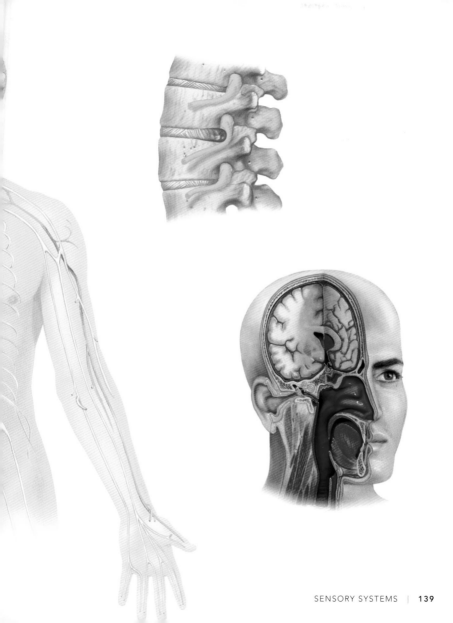

The sensory system

THE SENSORY SYSTEM OF THE HUMAN
BODY HAS A NUMBER OF RECEPTORS
THAT RESPOND TO DIFFERENT STIMULI.
These receptors will respond to internal
and external stimuli and convert these
into nerve impulses that inform the
nervous system of the body's
environment. The encoding of these
impulses is called sensory transduction.
There are specific sensory receptors,
and they are classified by the
environmental qualities to which they
are sensitive. These receptors are
chemoreceptors, mechanoreceptors,
nociceptors, photoreceptors, and
thermoreceptors. They may also be
classified according to the source of the
quality that they sense; therefore,
receptors may sense events that occur
away from the body as well as those in
the immediate external environment.
Additionally, there are internal
receptors sensing changes in blood
pressure, oxygen, and carbon dioxide
levels in the blood, and substances
released following tissue damage.
Other receptors sense our position in
space and the disposition of limbs,
which is very important in relation to
movement of the human body. These
are known as gravitational receptors
and proprioceptors.

For a response to occur, there is
usually one stimulus that the receptor
is most sensitive to, and this is known
as the adequate stimulus. For example,
vibration is the adequate stimulus for
receptors in the skin, which are called
Pacinian corpuscles. In addition,
thermoreceptors in the skin will be
stimulated by changes in temperature.

▶ **OVERVIEW OF THE SENSES**
Five key senses and the structures
involved in their control.

OVERVIEW OF THE SENSES

Balance
Receptors in the semicircular canals and otolith organs of the inner ear register position. Movements stimulate the hair cell receptors, which relay information to the brain via the vestibulocochlear nerve.

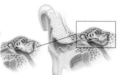

Hearing
The inner ear contains hair cell mechanoreceptors that relay signals along the vestibulocochlear nerve. The brain thus receives information on sound.

Smell
The tiny hairs (cilia) in the nasal cavity have chemoreceptors, which relay signals to the olfactory centers at the base of the brain, where they are translated to smells.

Sight
Photoreceptors in the eyes are sensitive to light. They send signals along the optic nerve, then to the brain's occipital cortex, which processes visual information.

Taste
The chemoreceptors on the tongue, throat, and palate are commonly known as taste buds. They transmit information on sweet, sour, salty, and bitter tastes to the brain via the cranial nerves.

The sensory system *(cont.)*

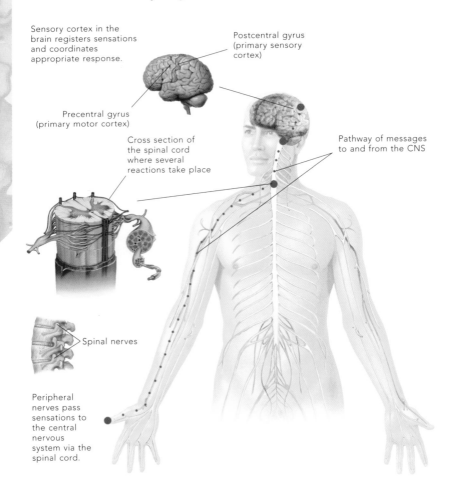

Sensory cortex in the brain registers sensations and coordinates appropriate response.

Postcentral gyrus (primary sensory cortex)

Precentral gyrus (primary motor cortex)

Cross section of the spinal cord where several reactions take place

Pathway of messages to and from the CNS

Spinal nerves

Peripheral nerves pass sensations to the central nervous system via the spinal cord.

◀ SENSORY PATHWAYS

The image depicts the various components of the sensory system and how the different nerve pathways link to the central nervous system.

The method by which the sensory system works is based on a feedback mechanism. A specific receptor reacts to the stimulus, sending an afferent signal to the central nervous system (CNS). In response, an efferent signal is sent back to the region involved, triggering an appropriate reaction. A number of receptors will be stimulated by a specific afferent nerve fiber, which has a receptive field. Nerve cells in the CNS can receive inputs from different afferent fibers, so the receptive field of one neuron in the spinal cord is usually larger than the afferent fibers connected to it—this is known as convergence. Following this, second-order nerve cells make contact with other nerve cells as information is processed, spreading sensory information among more nerve cells—this is known as divergence. Convergence and divergence are essential in the processing of sensory information in the body and are used to dictate patterns of behavior.

Proprioception

What is it?

PROPRIOCEPTION CAN BE DEFINED AS THE CONSCIOUS AND UNCONSCIOUS APPRECIATION OF JOINT POSITION. It is vital for joints to be able to function properly with unconscious control responsible for attuning muscle function and promoting joint stability. There are two mechanisms that work to coordinate the efferent responses known as feed-forward and feedback. Feed-forward refers to the planning of movements based on past experiences. Feedback occurs continuously to regulate motor control of the muscles acting around a joint through reflex pathways. Preparatory muscle activity occurs as a result of feed-forward, and reactive muscle activity as a result of feedback. The coordination of these two systems achieves dynamic restraint around a joint.

What affects it?

A RANGE OF MODIFIABLE AND NON-MODIFIABLE FACTORS CAN AFFECT PROPRIOCEPTION. An example of a nonmodifiable factor is age. As humans age, many of the body's systems slow down, including the proprioceptive feedback loop. Active individuals who exercise regularly and place their joints in positions of greater stress will have a training effect on the proprioceptors in those joints. Training status of an individual, therefore, can also have an effect on proprioception. In particular, traumatic injury can have a detrimental effect on proprioception. Therefore, an essential component of any rehabilitation program is proprioception and incorporating exercises that improve performance after sustaining an injury.

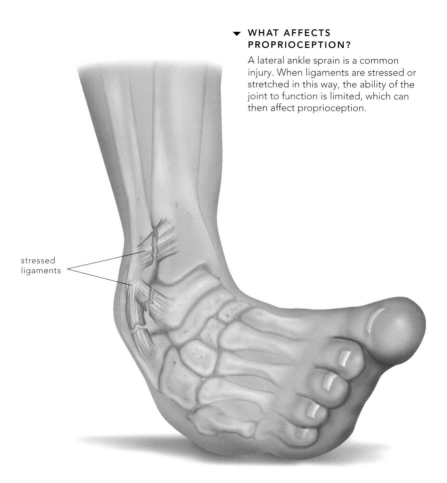

A lateral ankle sprain is a common injury. When ligaments are stressed or stretched in this way, the ability of the joint to function is limited, which can then affect proprioception.

stressed ligaments

Muscle spindles

MUSCLE SPINDLES ARE LOCATED WITHIN THE MUSCLES IN PARALLEL WITH MUSCLE FIBERS. Muscle spindles detect changes in the length of a muscle and, in particular, the rate of change of length. When a muscle spindle is stimulated, it transmits a signal via the afferent nerves to the central nervous system (CNS). Muscle spindles are innervated by small motor fibers called gamma-efferent nerves. This is important since the sensory and motor fibers are arranged independently, allowing for length changes to be accommodated while still transmitting afferent nerve signals. Once the signal reaches the CNS, it projects onto skeletal motor neurons through a monosynaptic reflex. This is a fast reaction, since only one synapse is involved. An efferent signal is then sent back to the muscle, thus causing a contraction of the muscle that was under stretch. This is known as the stretch reflex and is a protective reflex against injury.

Golgi tendon organs

GOLGI TENDON ORGANS ARE LOCATED IN THE TENDON AND THE TENDO-MUSCULAR JUNCTION. Their primary function is to detect tension in the muscle and tendon and cause a reflexive inhibition of the muscle. Golgi tendon organs are stimulated when a muscle is under load, and these receptors work to cause the opposite effect of muscle spindles. The feedback loop is the same as the arc that the muscle spindles work through in regard to the afferent signals sent to the CNS, where a monosynaptic reflex occurs and an efferent signal is sent back to the muscle, causing a reflex relaxation or inhibition. This is another protective reflex to help prevent injury.

▶ **MUSCLE SPINDLE AND GOLGI TENDON ORGAN**

The image shows the location and pathways of these receptors.

MUSCLE

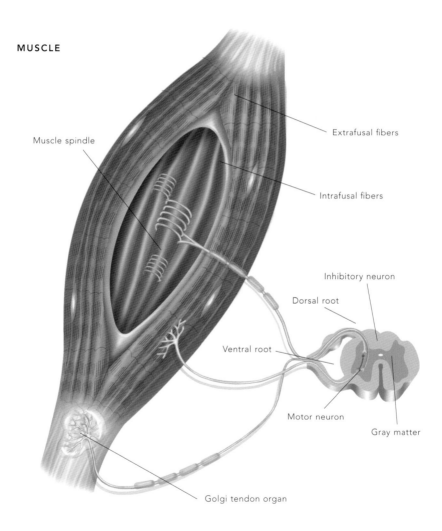

Muscle spindle

Extrafusal fibers

Intrafusal fibers

Inhibitory neuron

Dorsal root

Ventral root

Motor neuron

Gray matter

Golgi tendon organ

Joint receptors

LOCATED WITHIN JOINTS ARE ARTICULAR MECHANORECEPTORS. These receptors respond by reacting to tissue deformation that may occur around a joint. When the tissues around a joint are deformed, such as when injured, there is an increase in afferent discharge rate caused by the increased activation of mechanoreceptors. There are different types of mechanoreceptors; for example, in the knee there are

Pacinian corpuscles, Meissner corpuscles, and free nerve endings. Joint mechanoreceptors can be either quick-acting or slow-acting. Quick-acting receptors cease discharging shortly after the onset of the stimulus and provide conscious and unconscious kinesthetic sensations in response to joint movement. Slow-acting receptors continue to discharge while the stimulus is still present and provide

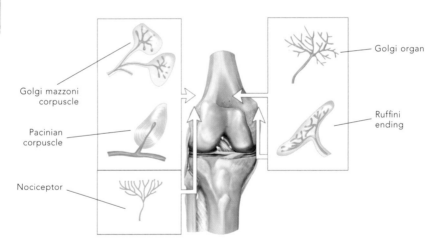

Golgi mazzoni corpuscle

Pacinian corpuscle

Nociceptor

Golgi organ

Ruffini ending

▲ **MECHANORECEPTORS**
The different mechanoreceptors of the knee.

continuous feedback regarding the position of the joint. Generally, joint mechanoreceptors only appear to be stimulated when under heavy load.

Following injury, such as a lateral ankle sprain, the joint mechano-receptors can be significantly disrupted, thus affecting the proprioception of the injured joint. However, with the correct input and training, the feedback system can be improved by helping the joint become more stable and helping to prevent further injury. With an injury such as this, evidence suggests that the lack of joint stability due to an absence of proprioceptive training is one of the biggest risks of reinjury, therefore demonstrating the importance of proprioception.

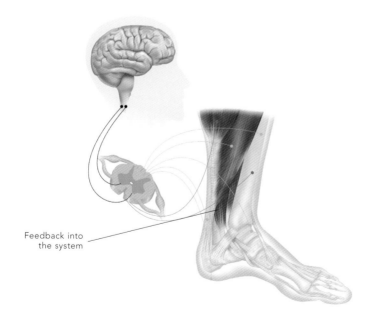

Feedback into the system

Reflexes

REFLEXES REPRESENT THE SIMPLEST FORM OF MOTOR ACTIVITY ELICITED BY THE NERVOUS SYSTEM. Each reflex includes at least two neurons: an afferent sensory neuron and an efferent motor neuron. Sensory afferent fibers transmit information from the receptor to the central nervous system (CNS) whereas efferent motor fibers relay information from the CNS to an effector muscle. The time taken between the stimulus and the response is called latency, and this can be affected by activity higher in the CNS.

The simplest reflex involves just two neurons and one synapse and is known as a monosynaptic reflex. Other reflex arcs may have a greater number of neurons between the afferent and efferent neurons, and these are called interneurons. One interneuron will mean that there are two synapses, and this reflex is therefore called disynaptic. If there are two interneurons, then three synapses are present, so the reflex is known as trisynaptic. Any greater numbers than these examples are known as polysynaptic. The

stretch reflex is monosynaptic, the withdrawal reflex is disynaptic, and the scratch reflex is polysynaptic.

The knee jerk is an example of a stretch reflex known as a myotatic reflex. When the quadriceps tendon is tapped, therefore lengthening it, the dynamic nuclear bag receptors of the muscle spindles are stimulated. There is then an increase in the rate of firing of the group Ia afferent fibers of those muscle spindles. When the afferent signal reaches the spinal cord, some of the signal makes monosynaptic contact with the alpha motor neurons that supply the quadriceps, causing a contraction, which abruptly extends the knee. The speed of the reflex will be affected when there is damage to the lumbar dorsal roots of the spinal cord. There are different reflexes that can be assessed depending on which level of the spine may be affected.

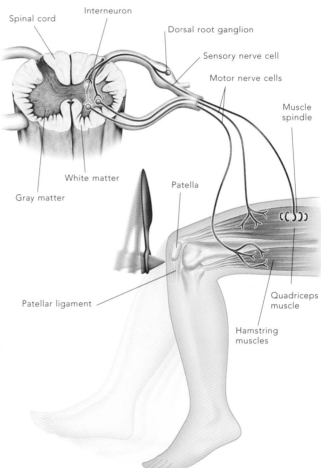

Spinal cord

Interneuron

Dorsal root ganglion

Sensory nerve cell

Motor nerve cells

Muscle spindle

White matter

Gray matter

Patella

Patellar ligament

Quadriceps muscle

Hamstring muscles

▶ **PATELLAR REFLEX**

An example of a myotatic reflex involving the patella tendon.

Sensory system in rehabilitation

Perhaps one of the best ways to demonstrate the sensory system in practice is by seeing how it can be incorporated into a rehabilitation program. A commonly used example of an injury that impairs the sensory system is the lateral ankle sprain. This injury is extremely common both in athletes and as a result of everyday activities.

Following a lateral ankle sprain, the ligaments will be damaged, the degree of which will depend on the severity of the injury and external forces placed upon it.

The injury will result in deformation of the tissues, which will stimulate the joint receptors. If the external load placed upon these tissues exceeds the individual's ability to withstand these forces, the result will be tissue damage. As a consequence, the ability of the proprioceptors to provide sufficient stability through the

▲ **REHABILITATION OF THE SENSORY SYSTEM**

An example of an advanced rehabilitation exercise that addresses proprioception. It works by retraining proprioceptors and increasing the stability they provide.

mechanisms previously outlined is compromised.

In order to regain stability, the proprioceptors need to be retrained. Initially, this will be done with less challenging tasks that will be progressed to more difficult challenges as the rehabilitation program advances.

Proprioceptive exercises should take place in a weight-bearing position in order for the joint to be loaded with the added compression, providing greater stability. An example of a straightforward proprioceptive exercise for ankle rehabilitation is a single-leg balance. This is often progressed by asking the individual to close his or her eyes, therefore removing visual feedback to the sensory system. This means that other parts of the system need to work harder to compensate for the lack of visual feedback; this includes the vestibular and somatosensory

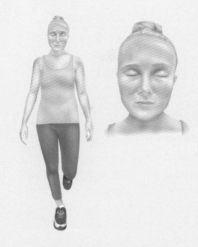

▲ **ADAPTING REHABILITATION**

By closing the eyes, one of the senses is removed from the system and the individual must rely more heavily on the other senses to remain balanced.

Sensory system in rehabilitation (cont.)

systems. There will be input from the proprioceptors and also from cutaneous receptors from the touch of the skin on the sole of the foot against the ground.

Further progression of exercises such as the single-leg balance would be to use equipment such as a wobble board. This increases the challenge on the individual to maintain stability without falling over. It will involve feedback from the joint receptors as well as the balance between the muscle spindles and Golgi tendon organs to maintain equilibrium. There will be influence through the feed-forward mechanism, as the earlier-stage exercises will have provided the past experiences that help to plan movements—and therefore the muscle activity that is required.

It should be noted that proprioceptive exercises should be included in any rehabilitation program. All athletes should also include this type of training into their normal routines to help maintain joint stability and potentially prevent injury.

REHAB AND EXERCISE ROUTINES

Proprioceptive exercises, such as rope-ladder activities, are integral to any comprehensive athletic exercise program.

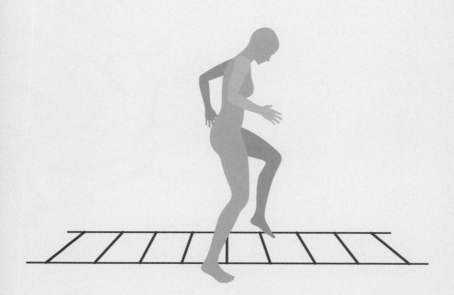

Chapter 7:
Supraspinal control of movement

The human brain is a highly complex organ, with an immense capacity to store and process information, while dictating the function of the human body. The brain consists of many different parts that work independently and together to control our lives. When things go wrong, conditions such as cerebral palsy and Parkinsons disease may develop, which have an adverse effect on the body's movement control.

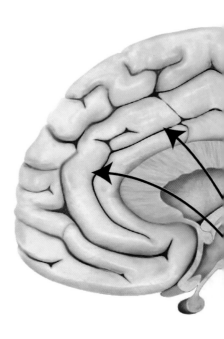

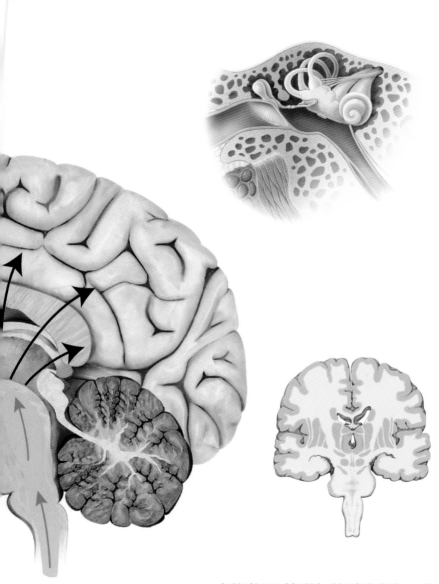

The cortex

THE BRAIN'S OUTERMOST LAYER IS TERMED THE CEREBRAL CORTEX. Constructed of gray matter, the cerebral cortex is roughly 5 millimeters in depth and encases the central axons of white matter, which form a large section of the brain. Although only a thin outer layer, the cerebral cortex makes up a substantial amount (roughly 40 percent) of the brain's total mass. The cerebral cortex is the most complex element of the nervous system and is constructed of a dense number of axons, all of which have a role in controlling the many functions of the body.

Visually, the cerebral cortex appears wrinkled due to a complex pattern of ridges, termed gyri (singular: gyrus) and furrows, termed sulci (singular: sulcus). This pattern of gyri and sulci allows for a large increase in the surface area of the brain. This increase in total surface area allows for a larger area of the brain to undertake complex processing tasks.

Within the cerebral cortex are also a number of deep grooves termed fissures, which offer separation between different segments. These individual sections are termed lobes and are named after the bones to which they sit adjacent. The cerebral cortex is also divided into two distinct cortices by a deep central fissure; this division in the sagittal plane allows for two sections, termed the left and right cerebral hemispheres. Although some primary elements of brain function are associated with a specific lobe or particular hemisphere, in general the segmentation of the cerebral cortex is more representative of anatomical variations than it is of differences in function. This is highlighted by elements of higher-level function, such as the cognitive processes involved in language construction, occurring in different lobes and opposing hemispheres of the cerebral cortex at the same time.

THE CEREBRAL CORTEX

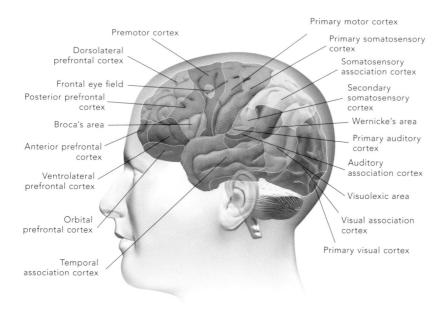

Premotor cortex

Dorsolateral prefrontal cortex

Frontal eye field

Posterior prefrontal cortex

Broca's area

Anterior prefrontal cortex

Ventrolateral prefrontal cortex

Orbital prefrontal cortex

Temporal association cortex

Primary motor cortex

Primary somatosensory cortex

Somatosensory association cortex

Secondary somatosensory cortex

Wernicke's area

Primary auditory cortex

Auditory association cortex

Visuolexic area

Visual association cortex

Primary visual cortex

▲ **CEREBRAL CORTEX**

The cerebral cortex can be divided into areas concerned with particular functions (e.g., vision, hearing, touch, and so on). These are usually consistent in position from one person to the next.

The cortex *(cont.)*

Functionally, the cortex may be considered to have two types, the primary and association cortices. The primary cortex is involved in the basic functions of sensory processing and motor output, while the association cortex allows for more complex processing tasks to take place. The term "association" is derived from the manner in which this functional subdivision of the cerebral cortex works. The association cortex considers a number of different types of information to complete more complex processing tasks. The basic functions of movement are controlled by the primary motor cortex; however, more complex maneuvers and processes, such as the planning of movement (consider sports visualization techniques), may require processing from information in the association cortex. Associated processing is also required to undertake the complex higher functions of the nervous system, such as cognition, emotion, and consciousness.

CROSS SECTION OF THE BRAIN

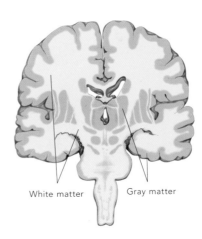

White matter   Gray matter

▶ **CEREBRAL ORGANIZATION**

The outermost layer of the brain is constructed from gray matter. This neural tissue has a pinkish-gray color and contains the cell bodies, dendrites, and axon terminals of neurons. Centrally, the brain is constructed from white matter, which consists of axons connecting different parts of gray matter to each other.

ORGANIZATION OF MOTOR AND SENSORY AREAS OF THE CEREBRAL CORTEX

Motor activity

Somatosensory activity

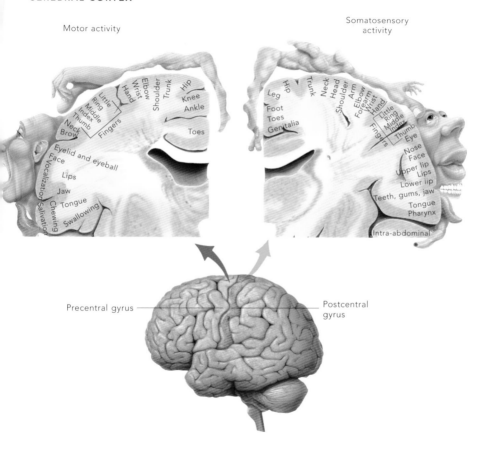

Precentral gyrus

Postcentral gyrus

Basal ganglia

LYING DEEP WITHIN THE BRAIN IS A CLUSTERED GROUP OF NUCLEI KNOWN AS THE BASAL GANGLIA. These neural structures function together, playing an important role in the control of movement. The basal ganglia group consists of four unique structures that control the ability to either promote or inhibit movement. The most superficial of these nuclei is termed the striatum—a long, tadpole-shaped structure that is made up of two distinct sections known as the putamen and the caudate nucleus.

Lying deep to the striatum is another cluster of neurons called the globus pallidus, which is functionally arranged into two subdivisions known as the internal and external globus pallidus. Two more structures, known as the subthalamic nucleus and substantia nigra, complete the basal ganglia complex.

The exact role of the basal ganglia in movement control is not yet fully understood. Current thoughts in this area support the concept of the direct and indirect pathways theory. Within this system, it is thought that the external globus pallidus plays an inhibitory effect on the thalamus, which in turn decreases output from the motor cortex. When the direct pathway is stimulated, neuronal activity in the cerebral cortex is received via the putamen, inhibiting the function of the external globus pallidus. This inhibition blocks the effect of the globus pallidus on the thalamus, which increases thalamic output back toward the cortex. The resultant increase in cortical activity triggers output from the motor cortex toward the body, initiating movement.

The indirect pathway offers an opposing system of movement regulation. Within this pathway, excitation of the subthalamic nucleus is thought to lead to increased suppression of thalamus activity, which in turn reduces the activity in the motor cortex, thus limiting movement.

▼ THE BASAL GANGLIA

The basal ganglia, including those shown below, are a clustered group of nuclei which lie centrally in the brain and play a role in the regulation of movement.

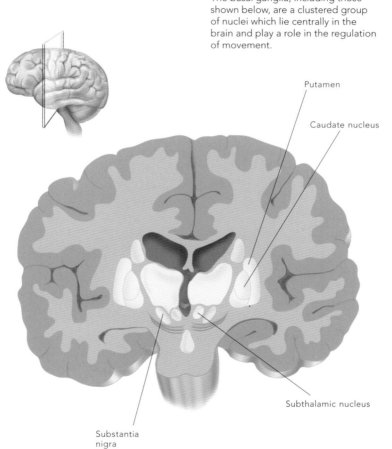

Putamen

Caudate nucleus

Subthalamic nucleus

Substantia nigra

Reticular formation

THE RETICULAR FORMATION IS A NONDESCRIPT COLLECTION OF NEURONS THAT LIE CENTRALLY WITHIN THE BRAINSTEM. This structure receives information from all major branches of the nervous system and has been suggested to play a role in filtering information to the sections of the brain that process sensory input and control movement. The reticular formation also plays a key role in channeling the output of the brain back toward the peripheral nervous system (PNS) via its descending pathway. This process of direct and indirect projection essentially offers the reticular formation a major role in all key functions of the nervous system.

Important roles of this section of the brainstem include control of autonomic functions, such as heart and breathing rate, alongside regulation of the tone in postural muscles.

The reticular formation receives afferent input in the form of sensory information from the PNS via the spinal cord. It couples this input with information gained from the vestibular and visual systems, which it receives from the brainstem and tectoreticular tract, respectively. This allows the reticular formation to filter incoming sensory input to the brain, allowing heightened or supressed sensations as required.

A key function the reticular formation plays in the regulation of movement is the control of extensor muscle tone. To overcome gravity, the body recruits the extensor muscles to maintain an upright posture. The resting level of tone required to maintain a standing posture is controlled in part by the function of the descending pathway. When movement is initiated via neural impulses in the cerebral cortex, signals pass from the motor cortex to the spinal cord via the reticular formation and this descending pathway. This process is excitatory to the extensor muscles and therefore leads to an increase in tone.

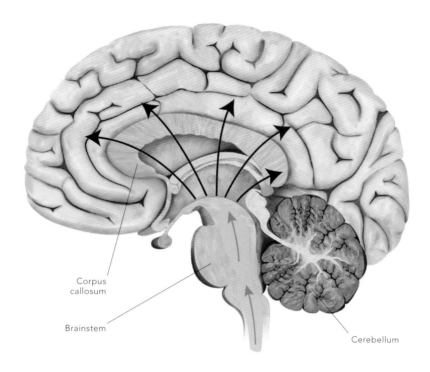

Corpus
callosum

Brainstem

Cerebellum

▲ **THE RETICULAR FORMATION**

The reticular formation is a projection of
the brainstem which filters information
(indicated here by the arrows) and helps
to regulate autonomic functions such as
heart and breathing rate alongside
control of tone in postural muscles.

Vestibular system

DEEP INSIDE THE INNER EAR, A CLUSTER OF AUDITORY RECEPTORS KNOWN AS THE COCHLEA PROCESSES INFORMATION ABOUT SOUND AND TRANSFERS NEURAL IMPULSES TO THE BRAIN. Adjacent to this spiral structure are three semicircular fluid-filled canals. These structures, coupled with another set of specialized sensory units known as the otolithic organs, make up the vestibular system. The body's vestibular function allows for information to be gained about the movement of the head, which in turn plays a key role in the maintenance of balance.

The semicircular vestibular loops, known individually as the anterior, posterior, and lateral (horizontal) canals, are each filled with a viscous fluid termed endolymph. The canals are orientated at right angles from one another, making each canal aligned along a different directional axis. Movement of the head causes increased flow of the endolymph through each of the semicircular canals. This allows sensory feedback to be gained regarding the movement of the head about each of the directional axes.

The otolithic organs consist of two discrete structures known as the utricle and saccule. These specialist sensory units allow for the sensation of acceleration and positional awareness of the head. This sensory function is enabled by a number of calcium carbonate crystals attached to small hairs, which are suspended in a viscous liquid inside the otolithic organs. The suspended nature of these crystals means that they are sensitive to forces such as gravity and acceleration. Since these forces act upon these structures, causing small movements and changes in position, action potentials are generated, causing nerve impulses to travel toward the brain and allowing for positional changes in the head to be interpreted.

VESTIBULOCOCHLEAR NERVE (VIII)

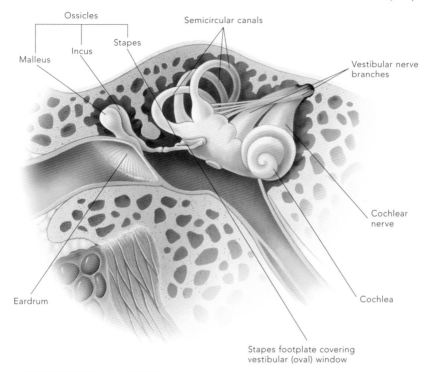

Ossicles

Malleus

Incus

Stapes

Semicircular canals

Vestibular nerve branches

Cochlear nerve

Cochlea

Eardrum

Stapes footplate covering vestibular (oval) window

▲ **THE VESTIBULAR SYSTEM**

Sensory information about motion, equilibrium, and spatial orientation is provided by the vestibular system. Inside the inner ear this specialized group of sensory organs detect movement of the head to aid balance.

Neurological impairments and disease

A number of conditions can affect the movement and motor control centers of the central nervous system. Each of these impairments can have serious effects on the ability to initiate and control movement.

Apraxia is a term given to a number of motor disorders that cause functional changes within the brain. Patients with this disorder demonstrate difficulty with conducting planned motor tasks. This means that they have trouble completing specific tasks, such as speech or movements, when instructed to do so. Apraxia usually affects areas of the brain that are related to higher-level function, such as the cerebral cortex. Dysfunction may also occur in the signaling pathways between different areas of the brain that control motor functions. The impairment is classified as an acquired condition, meaning it is usually caused by damage to the brain from injury, illness, or disease. The condition is often seen in patients who have suffered from cerebrovascular incidents such as stroke.

▶ **CROSS SECTION OF THE BRAIN**

The brain is central to the function of the central nervous system and acts as the control unit of the body. A number of illnesses, diseases, and impairments can affect this structure, leading to changes in the way the body moves.

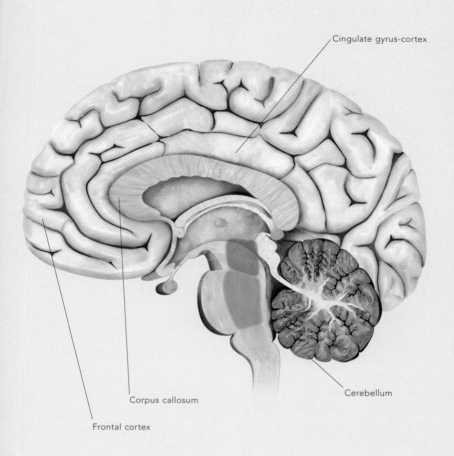

Cingulate gyrus-cortex

Corpus callosum

Cerebellum

Frontal cortex

Neurological impairments and disease *(cont.)*

Cerebral palsy (CP) is the term given to a number of different neurological conditions affecting areas of the brain that control movement, balance, and coordination. The condition is usually caused in fetal development but may also occur due to a traumatic birth or a head injury in early childhood.

Patients suffering from CP have altered motor control, demonstrating stiff and weak muscles. These alterations in motor function lead to changes in balance and coordination. The condition is usually classified by topographical distribution, meaning that classification is made to describe the body parts affected. For instance, "diplegia" describes when the legs are affected more than the arms. The term "hemiplegia" indicates that one side of the body is more affected than the other, while "quadriplegia" suggests changes in motor function are present in all four limbs.

▶ **CEREBRAL PALSY**

Cerebral palsy affects the central nervous system, causing loss of motor function in various regions of the body. Classifications can be made regarding the area of the body which is affected.

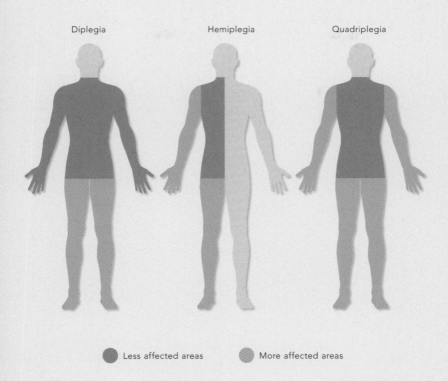

Diplegia   Hemiplegia   Quadriplegia

Less affected areas   More affected areas

SECTION THREE:
BIOMECHANICS

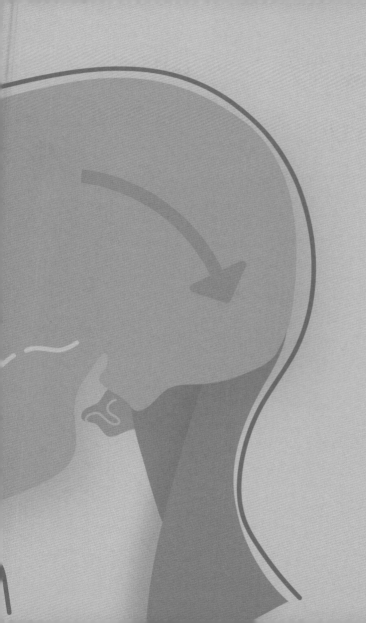

Chapter 8:
Principles of Movement

All human movement is activated by a force, which can take different forms and is influenced by many internal and external variables. The force needed to activate a movement depends on the mass upon which the force is exerted, with the limbs, joints, and muscles interacting to create the desired movement. External factors such as gravity and air resistance can vary the force needed to produce any given movement.

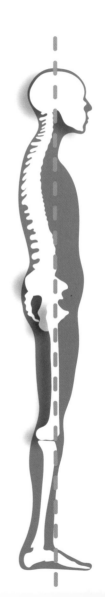

Force

What is force?

Biomechanics views force as the action caused on one body by another. Force is required for an object to move, stop, and change shape. The newton (N) is used as the unit to measure force, with one newton defined as the force required to accelerate one kilogram of mass at the rate of one meter per second squared. An understanding of forces and the actions they have on the body is essential when trying to understand human movement.

Kinematics and kinetics

The field of biomechanics concerns itself with two distinct areas known as kinematics—how the body moves—and kinetics—the forces by which movement is initiated. For example, the kinematic data obtained from a gait analysis may detail the angles of motion that occur at the knee joint at different time points throughout the gait cycle. Kinetic information from this type of analysis may be used to calculate the force generated from a muscle to control limb movements within gait. Understanding both the kinematics and kinetics of human movement can allow for better optimization of sporting movements. It can also allow clinicians to gain a greater understanding of how best to reduce and rehabilitate injuries.

FORCES INVOLVED IN RUNNING

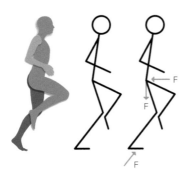

◀ **THE PHYSICS OF RUNNING**
Understanding the forces at work (indicated here by "F") during the gait cycle of a runner can help improve performance and reduce injury.

Newton's laws of motion

The current understanding of how forces act upon the human body is derived from Newtonian principles of physics. In 1687, Isaac Newton published the three-volume classic text, *Philosophiae Naturalis Principia Mathematica*, which detailed the three laws of motion alongside the universal law of gravitation.

Newton's First Law

"Every object persists in a state of rest or uniform motion unless it is compelled to change that state by forces impressed upon it."

Newton's first law is also known as the law of inertia and explains how an object will remain at rest unless a force is exerted upon it. A second consideration of this law is that once movement is initiated, an object will remain moving at the same velocity unless another force acts upon it. Within a sporting context, the first law of motion can be demonstrated by a penalty kick in soccer. Before the kick, the ball is in a state of rest; however, once the ball is struck, a force acts upon it and movement is initiated.

When considering human motion, it is important to consider that each movement must be initiated by a force, but it will also take a force to stop or change that movement.

Newton's Second Law

"The relationship between an object's mass (m), its acceleration (a), and its applied force (F) is F = ma. The direction of the force vector is the same as the direction of its acceleration vector."

Newton's second law describes how the rate of change in the velocity of an object (acceleration) is directly proportionate to its mass and the force applied. Consideration of the second law of motion is important when analyzing the forces involved in movement. Within a sporting context, the second law can be considered by a basketball player trying to make a shot from the three-point line with either a basketball or a bowling ball. Since the bowling ball has greater mass than the basketball, a greater force would have to be exerted to cause acceleration at the same rate toward the hoop. Using biomechanical analyses, knowing the mass of an object and its rate of acceleration can allow for calculation of the force that has been applied to initiate its movement.

THE LAWS OF MOTION

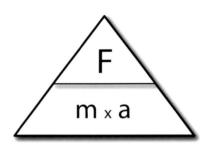

1. "A body will remain at rest or at constant velocity unless it is acted upon by an unbalanced force."

2. "The Force experienced by an object will be equal to its mass times acceleration" (F = ma).

3. "For every action there is an opposite and equal reaction."

▲ **CALCULATING FORCE**
Knowledge of an object's mass and its acceleration makes it possible to measure the force applied to it.

Newton's Third Law

"For every action there is an
 equal and opposite reaction."

The third law of motion states that
each action has an opposite and equal
reaction. This law can be demonstrated
by a tennis ball that is bounced on the
floor. The ball has mass and is
accelerating toward the floor, meaning
that it makes impact with a certain
amount of force. At the point of

contact, the ball is met with an
equal and opposite reaction force,
which dictates the rebound of the
ball. The equal and opposite reaction
from the floor is known as the ground
reaction force. This concept is
important in biomechanical analyses,
which considers the external forces
applied to the body throughout
human movement.

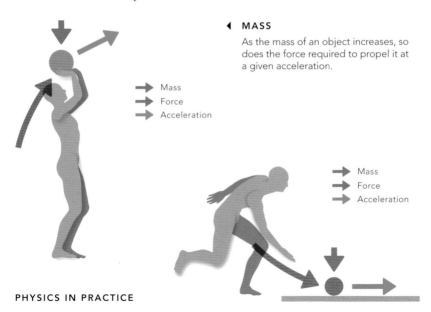

◀ **MASS**

As the mass of an object increases, so
does the force required to propel it at
a given acceleration.

➡ Mass
➡ Force
➡ Acceleration

➡ Mass
➡ Force
➡ Acceleration

PHYSICS IN PRACTICE

Forces

THE FIELD OF PHYSICS GENERALLY RECOGNIZES FOUR TYPES OF FORCES PRESENT IN NATURE: strong nuclear, weak nuclear, electromagnetic, and gravitational. When using biomechanical principles to consider human movement, the effect of gravity is of most concern.

"Two bodies attract each other with a force that is proportional to the product of their masses and inversely proportional to the square of the distance between them."

Gravity is an attraction force that occurs between any two objects that have mass. The more mass an object has, the greater its attraction force. When considering the forces involved in human movement, it is essential to consider the constant action that gravity has on the body. On or near to the Earth's surface, gravity is continually causing an accelerating effect on the body that is roughly equivalent to 9.81 m/s^2. At ground level this isn't always apparent, due to the

opposite and equal force being applied by the ground. When considering a skydiver, the effect of gravity is more apparent. When free-falling to Earth, a skydiver leaps from a plane and accelerates toward Earth, achieving a maximum speed (terminal velocity) of roughly 120 mph (200 kph). Gravity could accelerate the skydiver to greater speeds; however, the effect of other forces, such as air resistance, limits overall velocity.

Air resistance

Although mostly invisible to the human eye, the Earth's atmosphere is made of a dense array of gas particles. Movement in this environment means that the atmosphere provides a resistance force, which a moving body must overcome. This concept can be illustrated in sports such as running and cycling, where athletes commonly adapt their body positions to become more aerodynamic. By manipulating their body position, these athletes can become less susceptible to the force applied by air resistance. Much of this resistance comes from friction that occurs between the moving body and the gas particles that surround it.

Friction

Friction can be defined as the resistance to motion of two moving objects or surfaces that touch. Excessive friction is often viewed as detrimental and in some cases may be a driver for injury. However, the opposite can be true when considering many feats of sporting performance. Sprinters can use the friction generated between their foot and the running track to generate the large forces required to propel themselves forward at fast speeds. However, if they were to try this action in an environment with less friction—such as the surface of an ice rink—they would be unable to complete the task in the same manner.

Mass and weight

The mass of an object is the amount of matter an object is constructed of. When measuring the human form, weight—measured in newtons (N)—is regularly used; however, the truest unit of measure for mass is the kilogram (kg). The mass of an object is a constant, meaning that it doesn't change in different environments, whereas an object's weight is dependent on its mass and the acceleration effect that gravity has in that environment. On the surface of the Earth, gravity accelerates matter at 9.81 m/s^2, and this effect causes 1 kg of matter to weigh 9.81 N. In other environments, such as the surface of the moon, where gravity causes a lower level of acceleration, the same kilogram of matter would weigh less. Both the mass of the body and its subsequent weight are important factors for consideration in the analysis of human movement.

MOTION EQUATION

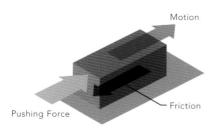

Motion

Friction

Pushing Force

Moment, moment arm, and torque

HUMAN MOVEMENT OCCURS DUE TO THE CONTRACTION OF MUSCLES, WHICH CAUSES MOTION AROUND VARIOUS JOINT AXES. The rotational force generated around an axis is termed torque. For torque to occur, both a moment of force and a moment arm must be present. The moment of force is defined as the application of a force at a perpendicular angle to a joint or point of rotation. The moment arm is the distance between the application of the moment of force and the axis of rotation. When considered in the human body, the axis of rotation is a joint, and the moment of force is the pulling force applied via a contracting muscle on a bone. The moment arm is the distance between the muscle's insertion and the joint's center of rotation.

Joint and muscle moment

The principles of moment of force, moment arm, and torque can be illustrated by considering the mechanics required to push open a door. If a pushing force is applied at the hinges, a door will not open. However, if the same pushing force is applied at the opposing end of the door, it will spin open around its fixed axis. This difference occurs due to the presence of a moment arm in the second scenario. As previously detailed, for torque to occur, a moment arm must be present. When pushing at the

▶ **TORQUE**

The greater the distance from the point of rotation to the moment of force (as indicated by the moment arm), the greater the torque.

TORQUE

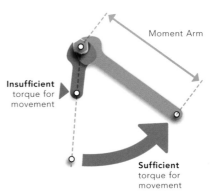

Moment Arm

Insufficient torque for movement

Sufficient torque for movement

hinge end of the door, there is no distance between the axis of rotation (hinges) and moment of force (push), meaning there is no moment arm.

When the force is applied at the other end of the door, a moment arm is present, meaning torque can be generated. When considering human movement, this offers some key considerations. A muscle's attachment must be at a sufficient distance (moment arm) from a joint's axis of rotation to generate movement and subsequent torque. Additionally, to initiate and control movement, muscles with an attachment located farther away from a joint are required to generate less force than muscles with a closer attachment. Understanding the locations of a joint's center of rotation and the attachment sites of muscles allows biomechanists to develop models that allow for the calculation of forces (moments) that are generated by different muscles throughout human movement.

MUSCLE MOMENT

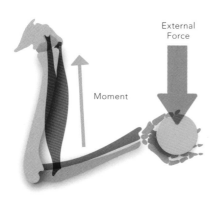

External Force

Moment

◀ **MUSCLE PROXIMITY**
The proximity of the muscle to the joint affects the force required to perform a particular function.

Pressure

THE TERM "PRESSURE" DESCRIBES THE FORCE ACTING OVER AN AREA. If a large force is applied over a large surface area, the relative pressure will be low.

Alternatively, if the same force is applied over a much smaller area, the pressure will be high. There is no universally accepted unit of measure for pressure; however, the unit most commonly used within biomechanics is derived from the pascal (Pa). One pascal is equal to the pressure present when one newton of force is applied over a surface area of one meter squared. In human movement, high pressures are experienced throughout the body, especially in areas such as the foot. The natural surface area of the body also makes the singular unit of a pascal not appropriate for use within biomechanical analyses. Therefore, the unit of kilopascal (kPa) is commonly utilized instead (1 kPa = 1,000 Pa).

PRESSURE

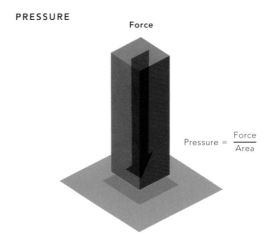

Force

$$Pressure = \frac{Force}{Area}$$

Center of pressure

The center of pressure is the point at which the total sum of pressure acts on a body, causing a force to act through that point. When considering a foot in contact with the floor, the center of pressure would be the point at which the ground reaction force is balanced on the plantar surface of the foot.

▼ WALKING

The shift in the center of pressure during five phases of a walking gait.

CENTER OF PRESSURE

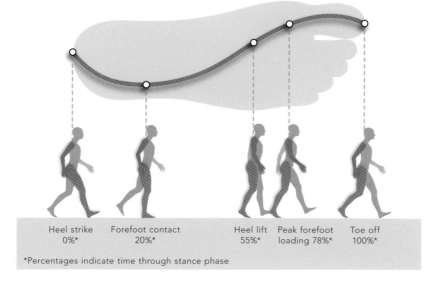

| Heel strike | Forefoot contact | | Heel lift | Peak forefoot | Toe off |
| 0%* | 20%* | | 55%* | loading 78%* | 100%* |

*Percentages indicate time through stance phase

Center of mass

The center of mass is a theoretical point on an object where its mass is evenly distributed. The center of mass is not always in the geometric center of an object; rather, it is the average location of that body's mass. The term "center of mass" is commonly interchanged with the term "center of gravity." Since the effect of gravity dictates the movement behavior of mass, this interchange in terminology is not completely accurate. However, within the Earth's atmosphere, the effect of gravity is a constant, meaning that the center of mass and center of gravity represent the same location on the body in this environment. Since most biomechanical analysis of human movement happens at ground level, this interchange in terminology is therefore deemed suitable.

Base of support

The term "base of support" considers the area enclosed by the points at which a body contacts a supporting surface. In human movement, this is commonly considered to be somewhere between both feet. The base of support is an important factor for consideration when trying to understand balance, stability, and equilibrium.

▶ **MASS DISTRIBUTION**

(Above) The base of support for an upright human is located at roughly the midway point between the two feet. (Below) The center of mass, shown here in terms of a figure moving through the stages of a somersault.

BASE OF SUPPORT

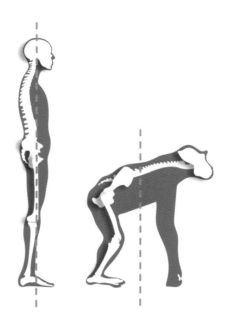

CENTER OF MASS/GRAVITY

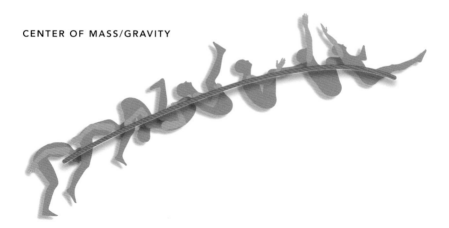

Balance and equilibrium

THE BODY'S RESISTANCE TO LINEAR AND ANGULAR FORCES IS TERMED STABILITY, OR EQUILIBRIUM. Balance is known as the ability to return the body to equilibrium after experiencing displacement from an external force. Balance and stability are vital when performing everyday tasks. Input from peripheral sources (eyes, vestibular organs, muscles, and joints) feeds into the brain, which integrates the information and sends a motor response to relevant muscles.

Factors that affect stability are location of the center of gravity (CoG), base of support, anticipation of the oncoming force, weight, and friction between contact surfaces. The base of support is the area between each point of support. For example, when standing, the base of support is small in comparison to a four-point position (hands and knees), which has a larger base of support. If the CoG falls within the base of support, then the body is stable. A stable body has a decreased likelihood of injury or a fall occurring when challenged by an external disturbance.

Equilibrium and stability can be categorized into three states: unstable, stable, and neutral. When an object is moved and continues to move, it is classified as unstable. For example, a ball on top of a sloped ramp that is displaced to either side will increase its displacement. A stable state of equilibrium is when an object is moved and returns to its original position. Neutral stability happens if an object is moved to a new position and remains in equilibrium.

▶ **BALANCE AND EQUILIBRIUM**
By moving through different positions on the beam yet maintaining balance, a gymnast achieves neutral stability.

EQUILIBRIUM

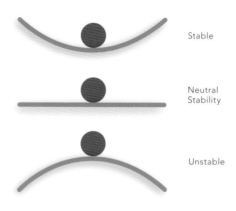

Stable

Neutral
Stability

Unstable

BALANCE

Levers

THE HUMAN BODY HAS EVOLVED OVER TIME INTO AN EFFICIENT ORGANISM, AND LEVERS EXEMPLIFY THIS EFFICIENCY. A lever is a simple machine that helps to make movement easier by magnifying the force and/or speed. Levers have four main components: a fulcrum (pivot point), a rigid bar, an effort force, and a resistance force. In the human body, these components are present as joints, bones, muscles, and body weight, respectively.

Levers help with movement by providing a mechanical advantage over the object being moved. Mechanical advantage (MA) quantifies how much easier the lever is making the work. It is expressed as the ratio of output to input. The higher the number, the easier the work is. An MA of one represents a neutral state in which effort and resistance are of the same value. A value of less than one means that the effort force is less than the resistance force; therefore, more effort is required to move the object, thus making the task harder.

There are three types of levers that can be found in the human body: first-class levers, second-class levers, and third-class levers.

A first-class lever is a lever that has the resistance and effort force on opposite sides of the fulcrum (e.g., a seesaw). First-class levers are uncommon within the human body, but an example of this type of lever is at the level of the first cervical vertebra and the skull. The weight of the head is the resistance force, the posterior muscles are the effort force, and the joint acts as the fulcrum. The MA of these levers is normally one.

▶ **THE THREE TYPES OF LEVERS**
The three classes of levers are all found in the human body.

FIRST CLASS

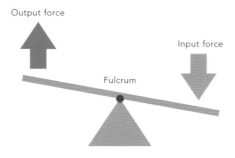

Output force

Input force

Fulcrum

SECOND CLASS

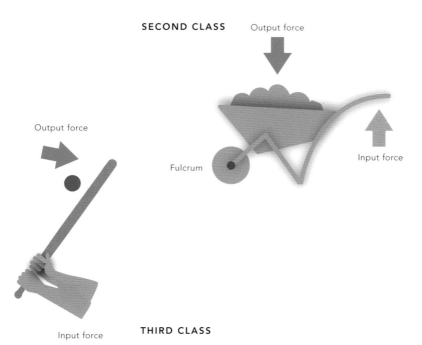

Output force

Fulcrum

Input force

Output force

Input force

THIRD CLASS

Levers *(cont.)*

A second-class lever has both the resistance and the effort force on the same side as each other, with the resistance force always acting in between the effecting force and the fulcrum (e.g., a wheelbarrow). These levers are rarely found in the human body, but one example that is regularly given is a heel raise. The metatarsophalangeal joints act as the fulcrum, with the effective force being applied via the muscles of the posterior lower leg to help overcome body weight (resistance). These levers often have an MA of more than one.

A third-class lever also has both forces on the same side of the fulcrum; however, the effective force is acting in between the fulcrum and the resistance force (e.g., a baseball bat or a shovel). This is the most common type of lever found in the human body. For example, consider the elbow joint: the resistance (weight) is farthest away, with the effort force (biceps) closest to the fulcrum (elbow joint). The MA of these levers is less than one.

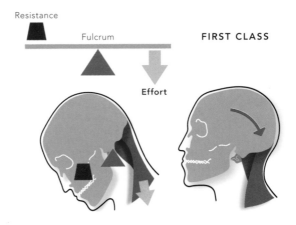

SECOND CLASS

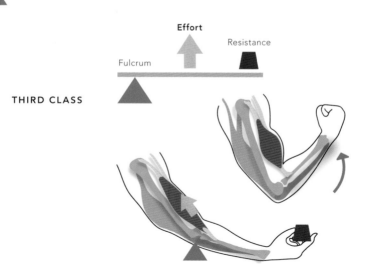

Third-class levers are more common in human anatomy, as shown below in the elbow joint. A first-class lever operates between the skull and the first cervical vertebra, while the heel is an example of the second class.

Effort

Resistance

Fulcrum

THIRD CLASS

Effort

Resistance

Fulcrum

Motion

Classification and practical applications

MOTION MAY BE REFERRED TO AS A CHANGE IN PLACE, POSITION, OR POSTURE THAT OCCURS OVER TIME AND IS RELATIVE TO A PARTICULAR POINT IN THE ENVIRONMENT. There are two types of motion: linear and angular.

Linear motion relates to movement along a straight or curved pathway where all points of the body move the same distance in the same amount of time. The direction, path, and speed of movement are the focus in these activities. The individual's center of mass (CoM) is most often the point in the body that is used for monitoring this motion. The CoM can be defined as the point on the body at which the distribution of mass is equal in all directions, and it does not depend on the gravitational field. However, if analyzing a particular skill—such as those in sports—other parts of the body may be used as the point to be monitored.

Angular motion refers to motion around a point in which different regions of the same body do not move the same distance in a set time. This often refers to motions around an axis, and a good sporting example would be swinging around a high bar in gymnastics. The gymnast's arms do not move as far as the legs. It is important to identify angular motions and the sequence in which they make up a skill, as they are often the key to successful completion of the desired movement.

▶ **MOTION**

(Above) Without outside forces, this object will never move.
(Below) Without outside forces, this object will never stop.

EXAMPLE 1

EXAMPLE 2

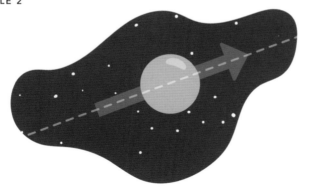

Acceleration and momentum

Acceleration refers to the change in velocity with respect to time. Acceleration is calculated as follows:

Acceleration = Change in velocity / change in time

The most commonly used unit of acceleration is meters per second. In human motion, the velocity of a body or segment is rarely constant. Velocity often changes, even if it appears that it remains the same. For example, a runner completing a training session of consecutive 400-meter intervals may appear to do each at the same velocity, but within that 400 meters, velocity will change and acceleration and deceleration will take place—although the athlete may not be aware of it. The terms "acceleration" and "deceleration" can be easily understood when an object is moving in the same direction. When there are changes in direction, the term "negative acceleration" may be used to indicate motion in the opposite direction, even with increasing velocity.

Momentum is the quantity of motion of an object, i.e., the relative amount of its motion, and is calculated by multiplying the mass of the body by the velocity. If a body is moving, it has momentum and will move in the direction in which a force is applied to it. When considering the components of momentum, it can be deduced that the heavier the object (or person), the greater the momentum. Angular momentum is the quantity of angular motion of an object and is classed as a vector measured in units of kilograms per meter squared per second. If gravity is the only external force acting on an object, the angular momentum remains constant throughout the duration of the movement. During walking there is forward, lateral, and vertical momentum, as there are forces acting on the body from different directions, i.e., gravity, ground reaction forces, and muscle forces.

▶ **THE RUNNER AND THE FIGURE SKATER**
(Above) After initial acceleration, the runner is unable to maintain the velocity and decreased acceleration is the result. (Below) Angular momentum is at play during a figure skater's tight spin.

ACCELERATION

ANGULAR MOMENTUM

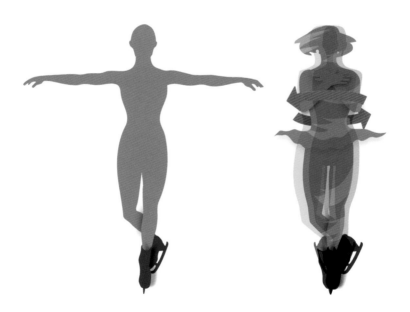

Velocity and displacement

VELOCITY RELATES TO HOW FAST A BODY IS MOVING, AND IT IS DETERMINED BY THE CHANGE IN DISPLACEMENT DIVIDED BY TIME. Displacement simply refers to the distance between the first position and the second position. Speed and velocity are terms that may be used interchangeably, but there are differences between them. Velocity is a vector quantity that describes magnitude and direction, whereas speed is a scalar quantity that describes magnitude only. The usual method to calculate velocity gives the average velocity over the time frame analyzed.

▶ **VECTOR VS SCALAR**

While scalar values refer only to the magnitude of weight (in this case, the weight of the bat lying on the ground), vectors refer to magnitude and direction (the weight of the bat moving toward—and striking—the ball).

▼ **VELOCITY**

As a vector quantity, velocity describes magnitude in a given direction. In this example, the direction is expressed in terms of angular displacement.

MEASURES OF SPEED AND VELOCITY

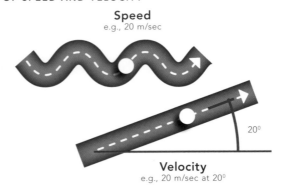

Speed
e.g., 20 m/sec

20^0

Velocity
e.g., 20 m/sec at 20^0

Vectors and scalars

IN BIOMECHANICS THERE IS A NEED TO DESCRIBE THE PHYSICAL QUANTITY OF WHAT IS BEING ANALYZED; therefore, its magnitude must be stated, and this can be done using scalars and vectors. A scalar is when you only need to describe the magnitude of something. An example may be the weight of an object, which refers to the magnitude of mass. If direction is needed also, then this is referred to as a vector, i.e., magnitude and direction. Force is an example of a vector, as it comprises magnitude and direction. When considering human movement, vectors often refer to movements of a limb or movements around a joint.

VECTOR QUALITIES

SCALAR QUALITIES

Sprinting and hamstring injury

Hamstring strains are one of the most common injuries among athletes and can occur in many sports, including sprinting. The most common point at which hamstring injuries occur is during deceleration of the limb in the swing phase of the gait, when the limb makes contact with the ground. When considering sprinters, one may also consider that their limbs may be further lengthened by the forefoot strike when running on a track in spikes. This may cause the sprinter to overreach even further, exposing them to a higher risk of injury.

At the point where most injuries occur, the hamstring muscle is maximally lengthened while being required to help slow down the limb through an eccentric contraction. The hamstring may fail, and fibers tear, if the muscle is unable to control the forces going through the limb at this point. The severity of the injury may be dependent on a number of factors, including the velocity at which the individual is traveling and the forces placed upon the limb, such as the ground reaction forces. It may also be affected by the force required to be exerted by the hamstring to control or slow down the limb. If other muscles are not working efficiently, the hamstrings may be required to compensate for this, adding stress to the muscle.

▶ **DECELERATION**

The hamstring muscle's ability to control the forces imposed upon it during deceleration will determine its susceptibility to damage.

SPRINTING

Sprinting and hamstring injury (cont.)

An accepted approach to hamstring rehabilitation and injury prevention is to train the hamstring muscle eccentrically. This increases the strength of the muscle in the more dangerous lengthened positions. It is important that the exercises selected are as closely related to the individual's activity as possible. For example, an individual who takes part in soccer will need to stress the hamstrings in multiple directions; therefore, the eccentric exercises should replicate those challenges. In the example of the sprinter, in which motion appears to occur in one plane of motion (100 meters), there are still external factors to be considered that may influence the athlete, such as the running track and the environment. Athletes in this sport may need to consider rehabilitation exercises that stress the hamstring with the hip in a fully flexed position and the ankle plantar flexed, in order to replicate the position where injury usually occurs.

Any rehabilitation program should consider the biomechanical influences of the sport in which the individual participates. Some of the principles considered in this chapter are easily integrated, such as creating better mechanical advantage through levers or by better controlling the external forces placed upon the athlete.

▶ **REHABILITATION OF THE HAMSTRING**

(Above) A sprinter tends to work in a single plane of motion and rehabilitation may focus too much on this plane.
(Below) An example of an exercise that loads the hamstrings eccentrically.

THE SPRINTER

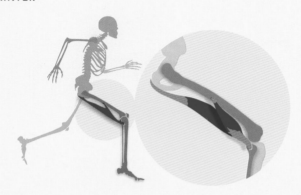

HIP FLEXION

SECTION FOUR:
ANATOMY OF MOVEMENT

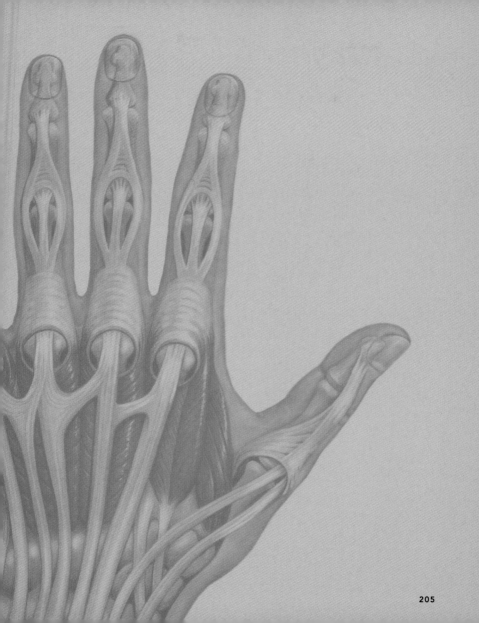

Chapter 9:
Shoulder girdle and upper limb

The shoulder and upper limbs, or arms, are a complex series of bones, muscles, tendons, and ligaments. Bones vary in size and nature, and joints allow for actions that include pushing, punching, pulling, and throwing. The shoulder and arms combine to produce many types of movement, including throwing and punching.

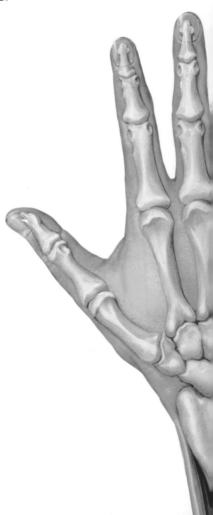

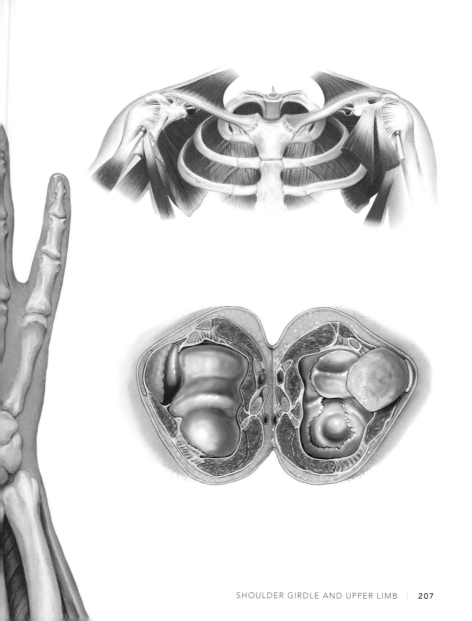

Bones of the arm

Bones of the upper arm

ANATOMICALLY, THE ARM IS THE SECTION OF THE UPPER LIMB BETWEEN THE SHOULDER AND THE ELBOW JOINTS. The humerus is the only bone in the arm, and it proximally articulates with the scapula at an angle of approximately 135 degrees to the axis of the arm. Below the head of the humerus is a narrow anatomical neck and two prominent bony ridges called tubercles. The greater and lesser tubercle—and the intertubercular groove that runs between them—provides an important attachment site for muscles that move and stabilize the arm.

On the lateral side of the humeral shaft is a roughened surface where the deltoid muscle inserts. This is known is as the deltoid tuberosity. Distally, the shaft changes from a cylindrical shape to a broader, flatter surface. Distally, the humerus has two rounded articular areas called the medial and lateral epicondyles. The medial epicondyle is the larger of the two and provides a site for attachment of many of the flexor muscles of the forearm. The lateral epicondyle, however, is the origin of many extensor muscles of the forearm. The distal humerus also has three indentations, or fossa, to allow for movement of the bones of the forearm during flexion and extension.

▸ **THE UPPER ARM**
The humerus articulates at one end with the scapula and at the other with the radius and ulna.

BONES OF THE ARM

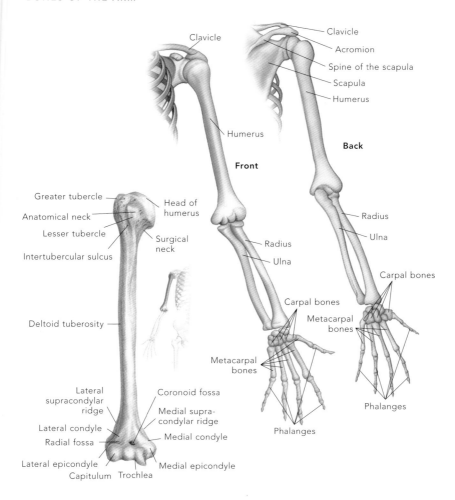

Clavicle

Clavicle

Acromion

Spine of the scapula

Scapula

Humerus

Humerus

Back

Front

Greater tubercle

Head of humerus

Anatomical neck

Lesser tubercle

Surgical neck

Intertubercular sulcus

Radius

Ulna

Radius

Ulna

Carpal bones

Deltoid tuberosity

Carpal bones

Metacarpal bones

Lateral supracondylar ridge

Coronoid fossa

Medial supra-condylar ridge

Metacarpal bones

Lateral condyle

Radial fossa

Medial condyle

Phalanges

Lateral epicondyle

Medial epicondyle

Phalanges

Capitulum   Trochlea

Bones of the forearm

THE FOREARM CONTAINS TWO LONG BONES CALLED THE RADIUS AND THE ULNA, WHICH RUN PARALLEL TO EACH OTHER. The longer and larger of the two bones is the ulna, which is on the medial side of the forearm in the anatomical position. Proximally, the ulna has a c-shaped articular surface called a trochlear notch that fits over the trochlea of the humerus to form the hinge of the elbow joint. The olecranon process of the ulna hooks around the distal humerus and forms the bony tip of the elbow. The shaft of the ulna is triangular in shape (i.e., in section) and narrows considerably distally, where it articulates with the radius.

Located on the lateral side of the forearm, the radius is the shorter and thinner of the forearm bones. The radius is narrowest at the elbow and widens as it extends distally. Proximally, the radius has a cylindrical head that articulates with the radial notch of the ulnar and forms the proximal radioulnar joint, permitting rotation of the lower arm and hand. Distally, the radial shaft expands to form a rectangular end. On the medial surface, the ulnar notch articulates with the head of the ulna and forms the distal radioulnar joint. While the ulna acts as the stabilizing bone, the distal end of the radius rotates around the ulna when the hand and forearm pronate and supinate.

▶ **FOREARM**

At one end, the radius and ulna articulate with the humerus at the elbow joint. At the other end, they form the wrist.

BONES OF THE FOREARM

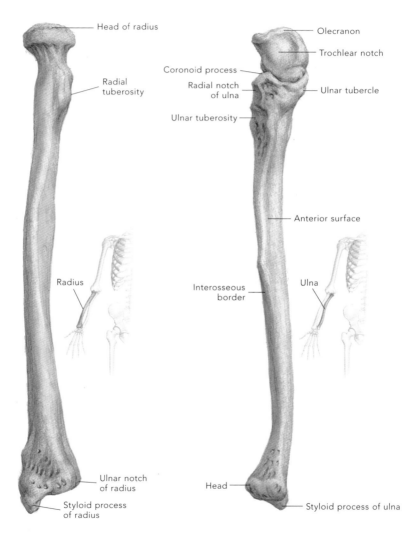

Head of radius

Radial
tuberosity

Radius

Ulnar notch
of radius

Styloid process
of radius

Coronoid process

Radial notch
of ulna

Ulnar tuberosity

Interosseous
border

Olecranon

Trochlear notch

Ulnar tubercle

Anterior surface

Ulna

Head

Styloid process of ulna

Arm muscles

Deep muscles of the arm

MANY OF THE MUSCULAR MOVEMENTS OF THE ARM ORIGINATE IN THE SHOULDER; however, there are a number of deep arm muscles that contribute to movement of the shoulder and elbow joint.

On the anterior side of the arm, the brachialis lies deep to the biceps brachii. The brachialis originates on the anterior humerus and inserts on the proximal ulna. This muscle functions to flex the elbow and is the only muscle that will perform this function from every arm position. Superficial to this is the biceps brachii, which has two origins on the scapula. The short head of the biceps brachii originates at the coracoid process at the top of the scapula. The long head originates at the supraglenoid tubercle, just above the shoulder joint. The common flexor origin on the medial condyle is visible, and this is the origin for a number of the muscles of the forearm.

On the posterior aspect, the triceps brachii is the only muscle on the arm. The triceps brachii is a three-headed muscle that distally converges to a common tendon. Superiorly, its origins are visible, with the larger long head arising from the infraglenoid tubercle of the scapula. The lateral head arises from the greater tubercle on the proximal humerus. The medial head, mostly covered by the long and lateral heads, arises from the radial groove of the humerus on the posterior midshaft.

▶ **WRIST MOVEMENT**

The muscular action of the flexor carpi radialis, flexor carpi ulnaris, and palmaris longus flexes the wrist, while the extensor muscles of extensor carpi ulnaris and extensor carpi radialis both extend the wrist.

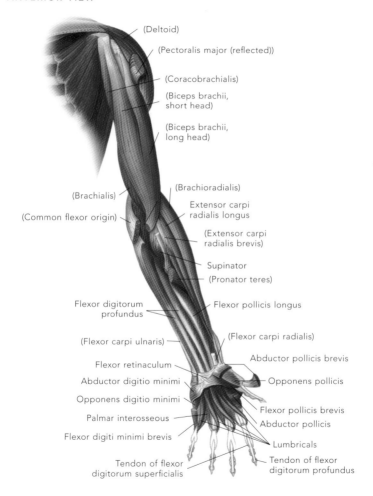

(Deltoid)

(Pectoralis major (reflected))

(Coracobrachialis)

(Biceps brachii, short head)

(Biceps brachii, long head)

(Brachioradialis)

(Brachialis)

Extensor carpi radialis longus

(Common flexor origin)

(Extensor carpi radialis brevis)

Supinator

(Pronator teres)

Flexor digitorum profundus

Flexor pollicis longus

(Flexor carpi ulnaris)

(Flexor carpi radialis)

Flexor retinaculum

Abductor pollicis brevis

Abductor digitio minimi

Opponens pollicis

Opponens digitio minimi

Palmar interosseous

Flexor pollicis brevis

Abductor pollicis

Flexor digiti minimi brevis

Lumbricals

Tendon of flexor digitorum superficialis

Tendon of flexor digitorum profundus

Surface muscles of the arm

THE SUPERFICIAL MUSCULATURE OF THE
ARM GIVES THE ARM ITS SHAPE, AND
THE CONTOURS IT FORMS ARE OFTEN
VISIBLE ON THE SURFACE. The belly of
the biceps brachii muscle lies anteriorly
to all other musculature on the anterior
arm. Distally, it attaches to the radial
tuberosity on the radius, allowing it to
supinate the forearm and flex the
elbow simultaneously. It also inserts
distally into the fascia tissue via the
bicipital aponeurosis. The tough
aponeurosis provides protection for the
median nerve and brachial artery that
run beneath it.

In the posterior compartment, the
triceps brachii is covered proximally
near its origin by the posterior fibers of
the deltoid muscle. Distally, the three
heads converge to a common tendon,
which inserts onto the olecranon
process of the ulna and is continuous
with fascia extending into the forearm.
The triceps brachii is the most
important extensor of the elbows, but
as it crosses the shoulder joint, it also
contributes to adduction and extension
of the arm.

▶ **MOVEMENT AT THE ELBOW**
Many of the muscular movements
of the arm originate in the shoulder.
The agonist and antagonist pairing
of the biceps and triceps results in
bending and straightening of the
arm at the elbow.

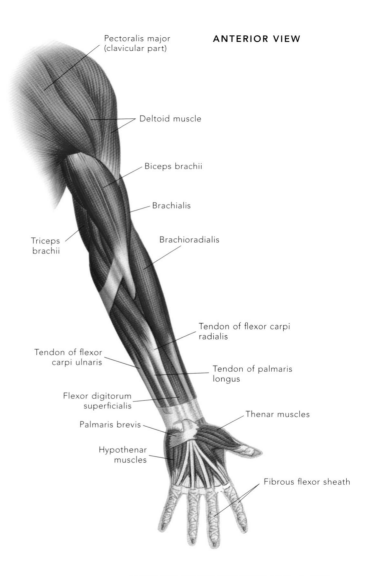

Pectoralis major
(clavicular part)

Deltoid muscle

Biceps brachii

Brachialis

Brachioradialis

Triceps
brachii

Tendon of flexor carpi
radialis

Tendon of flexor
carpi ulnaris

Tendon of palmaris
longus

Flexor digitorum
superficialis

Thenar muscles

Palmaris brevis

Hypothenar
muscles

Fibrous flexor sheath

Bones of the shoulder

Anterior view of the shoulder

THE SHOULDER GIRDLE IS COMPOSED OF A SET OF THREE ARTICULATIONS, WHICH INVOLVE THE STERNUM, CLAVICLE, SCAPULA, AND HUMERUS. The clavicle, or collarbone, is a long bone with an s-shaped curve that extends from the sternum to the scapula. The medial end has a large facet to articulate with the sternum. The sternoclavicular joint is the articulation between the medial clavicle and the clavicular facet on the upper lateral sternum. It forms the main link, and the only bony connection, between the upper limb and the axial skeleton. The shaft of the clavicle is the site for the attachment of a number of muscles in the region. The lateral end has a facet for articulation with the acromion of the scapula, forming the acromioclavicular joint.

The clavicle, along with the scapula, forms the pectoral girdle, attaching the bones of the arm to the trunk. The clavicle anchors the arms to the trunk while permitting the movement of the scapula and shoulder joints relative to the trunk. The movement of the clavicle increases the mobility of the shoulder joints. The scapula, or shoulder blade, is a triangular and flat bone that serves as a site for attachment for a number of muscles. It articulates with the humerus at the glenohumeral joint, and with the clavicle at the acromioclavicular joint.

▶ **MOVEMENTS OF THE SHOULDER**

Generally, muscles that pass in front of the shoulder joint act to flex or rotate the humerus. Those that pass behind the shoulder extend and/or rotate the humerus. The muscles passing over the shoulder abduct the humerus.

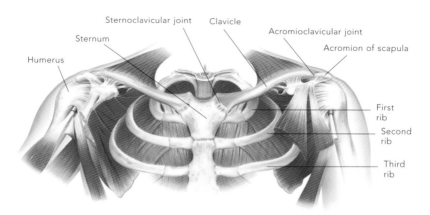

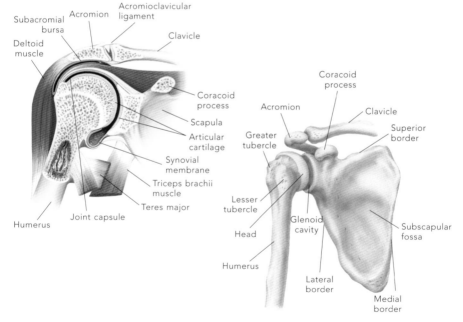

Posterior view of the shoulder

THE ANTERIOR ASPECT OF THE SCAPULA IS CALLED THE COSTAL SURFACE, which has a concave depression called the subscapular fossa and is the site of attachment for the subscapularis muscle. Superiorly and laterally to this is the coracoid process, which is a hook-like projection that lies just underneath the clavicle and provides the site for the attachment of three muscles.

The majority of the rotator cuff muscles, which stabilize the shoulder joint, attach to the posterior aspect of the scapula. A prominent feature of the posterior scapula is the spine, which traverses the scapula. Below the spine is the infraspinatous fossa, and above is the supraspinatous fossa—both of which provide sites for more rotator cuff muscle attachments. The acromion process is a lateral projection of the spine that articulates with the clavicle to form the acromioclavicular joint.

Below the acromion process is the glenoid fossa, a cavity that articulates with the humerus to form the glenohumeral, or shoulder, joint. The head of the humerus is much larger than the shallow concavity of the glenoid fossa, thus giving the joint an inherent instability. However, the glenoid fossa is deepened by a fibrocartilage rim, called the glenoid labrum, which, in conjunction with the extensive musculature around the joint, provides stability.

▶ **THE SCAPULA**

The scapula articulates with the humerus to create the shoulder joint. It is also attached to the clavicle and humerus.

SHOULDER

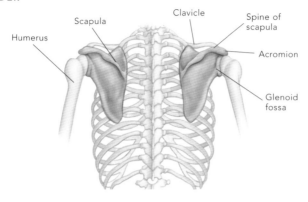

Humerus
Scapula
Clavicle
Spine of scapula
Acromion
Glenoid fossa

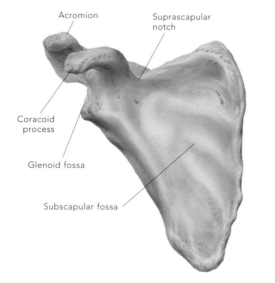

Acromion
Suprascapular notch
Coracoid process
Glenoid fossa
Subscapular fossa

Ligaments and tendons of the shoulder

THE STERNOCLAVICULAR JOINT IS HELD IN PLACE BY A NUMBER OF CAPSULAR LIGAMENTS. The anterior and posterior sternoclavicular ligaments reinforce the joint capsule anteriorly and posteriorly. The costoclavicular ligament attaches at the first rib inferiorly, and to the anterior and posterior borders of the clavicle superiorly, and stops the clavicle from being pulled away from the thorax. The interclavicular ligament connects the sternal ends of each clavicle and strengthens the joint capsule superiorly.

The acromioclavicular joint capsule and ligaments surrounding the joint work together to provide stability and maintain the articulation between the clavicle and the acromion process of the scapula. There are three major ligaments present in this joint. The acromioclavicular ligament spans from the acromion to the lateral clavicle, covering the joint capsule and reinforcing its superior aspect. Collectively, the conoid and trapezoid ligaments are known as the coracoclavicular ligament. The conoid ligament runs vertically between the coracoid process of the scapula to the conoid tubercle of the clavicle. The trapezoid ligament connects the coracoid process of the scapula to the clavicle.

The joint capsule and the ligaments of the glenohumeral joint provide support to keep the humeral head in contact with the glenoid fossa. The coracoacromial ligament runs between the acromion and the coracoid process of the scapula. It prevents superior displacement of the humeral head. The other ligaments of the shoulder joint can be considered thickenings of the joint capsule. The superior, middle,

▶ **THE SHOULDER JOINT**
The ligaments and tendons of the shoulder surround the joint, facilitating movement by maintaining articulation and ensuring stability.

and inferior glenohumeral ligaments are three bands that run from the glenoid fossa to the anatomical neck of the humerus and stabilize the anterior aspect of the joint. The coracohumeral ligament attaches to the base of the coracoid and runs to the greater tubercle of the humerus and supports the superior part of the joint capsule. The transverse humeral ligament covers the intertubercular groove of the humerus and holds the tendon of the long head of the biceps in place.

POSTERIOR VIEW

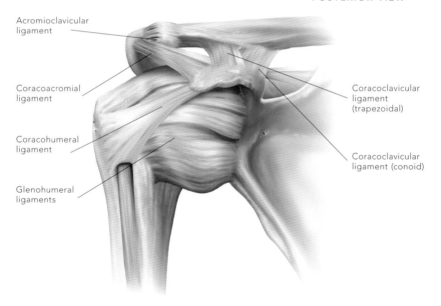

Acromioclavicular ligament

Coracoacromial ligament

Coracohumeral ligament

Glenohumeral ligaments

Coracoclavicular ligament (trapezoidal)

Coracoclavicular ligament (conoid)

Muscles of the shoulder

Anterior view of the shoulder

THE PECTORALIS MAJOR IS A LARGE AND PROMINENT FAN-SHAPED MUSCLE WITH A STERNAL HEAD ORIGINATING ON THE STERNUM AND COSTAL CARTILAGE OF RIBS 1–6 AND A CLAVICULAR HEAD ORIGINATING ON THE MEDIAL HALF OF THE CLAVICLE. It inserts on the lateral lip of the intertubercular groove of the humerus. The pectoralis major is an adductor and medial rotator of the arm, while the clavicular head can also contribute to shoulder flexion.

Deep to the pectoralis major muscle is the pectoralis minor. It is a smaller muscle and acts on the scapula rather than the arm. It depresses, or pulls down, the scapula and draws it toward the thoracic wall. It originates on ribs 3–5 and inserts on the medial border of the coracoid process of the scapula. Lateral to the pectoralis minor is the serratus anterior, which originates on ribs 2–8 and inserts on the medial border of the scapula. The serrate minor draws the scapula forward and upward, thus rotating it and allowing a greater range of motion of the shoulder.

The trapezius is a large muscle with fibers running in different directions. It originates at the occipital protuberance of the skull, the spinous processes of the vertebrae between C7 and T12 and the ligamentum nuchae, which joins the spinous processes. The fibers insert on the clavicle, the acromion, and the spine of the scapula. The upper fibers of the trapezius elevate the scapula and rotate it during abduction of the arm. The middle fibers retract the scapula, pulling it toward the spine, while the lower fibers pull the scapula inferiorly.

The biceps brachii also contributes to shoulder flexion through the long head attaching on the supraglenoid tubercle and the short head on the coracoid process of the scapula. In addition, the coracobrachialis, which is deep to the biceps brachii, originates at the coracoid process and inserts on the medial border of the mid-humerus, assisting in arm flexion.

The latissimus dorsi is a large muscle that originates in the lower back and covers a wide area. It originates at the spinous processes of T6–T12, the iliac crest of the pelvis, the thoracolumbar fascia, and the lower three ribs. It converges into a tendon that attaches to the intertubercular groove of the humerus. It powerfully extends, adducts, and medially rotates the arm.

Teres major is a thick, flattened muscle that adducts and extends the arm. It originates on the lower lateral border and inferior angle of the scapula and inserts on the medial lip of the intertubercular groove of the humerus.

▼ **SHOULDER MUSCLES**

Muscles attaching the humerus to the shoulder girdle include the deltoid, pectoralis major, latissimus dorsi, teres major, and the rotator cuff muscles.

PECTORAL GIRDLE

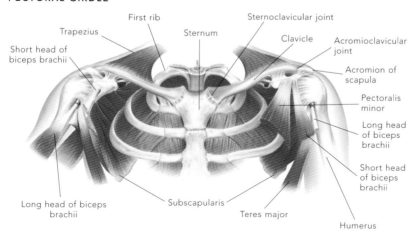

First rib
Trapezius
Sternum
Sternoclavicular joint
Clavicle
Acromioclavicular joint
Acromion of scapula
Short head of biceps brachii
Pectoralis minor
Long head of biceps brachii
Short head of biceps brachii
Long head of biceps brachii
Subscapularis
Teres major
Humerus

Posterior view of the shoulder

THE MOST PROMINENT MUSCLE OF THE SHOULDER REGION IS THE DELTOID MUSCLE. It is v-shaped, or like an upside down Greek letter delta (Δ), giving it its name. It originates from the lateral end of the clavicle, the acromion of the scapula at the top of the shoulder and the spine of the scapula, and converges to attach to the deltoid tuberosity on the lateral surface of the humerus. The anterior fibers flex the arm at the shoulder, while the posterior fibers extend the arm at the shoulder and the middle fibers are the major abductor of the arm. Although it is the major extensor of the arm, the triceps brachii also acts on the shoulder. With its long head originating at the infraglenoid tubercle of the scapula, it contributes to shoulder extension.

In order to maintain stability of the shoulder, a group of muscles work together to pull the humerus into the glenoid fossa. Collectively, they work together to limit excessive sliding movement of the humerus on the glenoid, but individually, they each have their own actions. These muscles are known as the rotator cuff muscles and include the subscapularis, infraspinatus, teres minor, and supraspinatus.

The subscapularis medially rotates the arm and originates at the subscapular fossa and inserts on the lesser tubercle of the humerus. The infraspinatus is the main lateral rotator of the arm; its origin is below the spine of the scapula in the infraspinous fossa, and it inserts on the greater tubercle of the humerus. The teres minor also laterally rotates the arm and assists with adduction. It originates on the middle part of the lateral border of the scapula and inserts at the inferior aspect of the greater tubercle of the humerus. The supraspinatus, however, has no specific action of its own but plays an important role in stabilizing the glenohumeral joint by pulling the humerus toward the glenoid, particularly during abduction. Its origin is above the spine of the scapula in the supraspinous fossa, and it inserts on the greater tubercle of the humerus.

ROTATOR CUFF MUSCLES

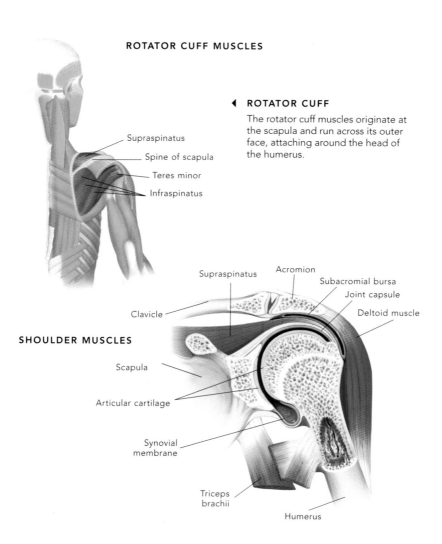

◀ ROTATOR CUFF

The rotator cuff muscles originate at the scapula and run across its outer face, attaching around the head of the humerus.

Supraspinatus

Spine of scapula

Teres minor

Infraspinatus

Supraspinatus

Acromion

Subacromial bursa

Joint capsule

Clavicle

Deltoid muscle

SHOULDER MUSCLES

Scapula

Articular cartilage

Synovial membrane

Triceps brachii

Humerus

Bones of the elbow

THE ELBOW JOINT IS A HINGE JOINT FORMED BY THE ARTICULATION BETWEEN THE DISTAL HUMERUS AND THE PROXIMAL ULNA AND RADIUS. The articulation between the ulna and humerus, and between the radius and humerus, constitutes the elbow joint, while the articulation between the radius and ulna is a pivot joint that allows rotation of the radius.

The distal humerus is shaped to accommodate the articulation of the forearm bones and to allow their range of movement around the joint. The distal humerus has two joint processes: the trochlea on the medial side and the capitulum on the lateral side. The trochlea articulates with the trochlear notch on the ulna. On the lateral side, the rounded capitulum articulates with the radius. The humerus has small indentations on the anterior aspect; the radial fossa and the coronoid fossa allow the humerus to accommodate the head of the radius and the coronoid process of the ulna when the elbow is fully flexed.

On the proximal ulna, the trochlear notch forms the hinge of the elbow joint, which is carved out of the olecranon process. The olecranon posteriorly acts as an insertion point for muscles crossing the elbow joint. At the proximal radius, the superior aspect of the head of the radius forms part of the hinge portion of the elbow joint, while the medial circumferential aspect of the head of the radius forms the radioulnar joint with the ulna.

▶ **THE ELBOW**

The elbow joint is formed by the articulation of the humerus, radius, and ulna. It consists of two joint types, a hinge joint and a pivot joint.

BONES OF THE ELBOW

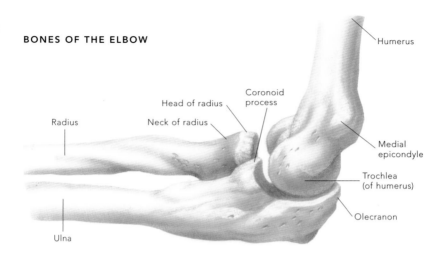

Humerus

Coronoid process

Head of radius

Neck of radius

Radius

Medial epicondyle

Trochlea (of humerus)

Olecranon

Ulna

INSIDE THE ELBOW JOINT

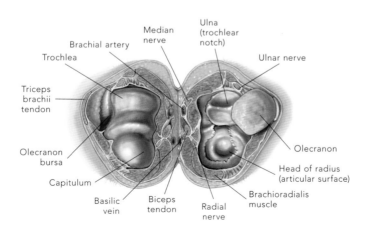

Median nerve

Ulna (trochlear notch)

Brachial artery

Trochlea

Ulnar nerve

Triceps brachii tendon

Olecranon bursa

Olecranon

Capitulum

Head of radius (articular surface)

Basilic vein

Biceps tendon

Radial nerve

Brachioradialis muscle

Ligaments and tendons of the elbow

AS WITH ALL SYNOVIAL JOINTS, THE ELBOW JOINT HAS A CAPSULE ENCLOSING THE JOINT. This thick, fibrous capsule has thickenings that form the collateral ligaments. These help prevent movement medially and laterally during flexion and extension.

On the medial side of the joint, the ulnar collateral ligament prevents lateral deviation of the joint. It consists of two triangular bands, anterior and posterior. It spans from the medial epicondyle of the distal humerus to the coronoid and olecranon processes of the ulna. The radial collateral ligament prevents medial deviation at the elbow joint. It is a short, narrow band that extends from the lateral epicondyle of the humerus to the annular ligament on the radioulnar joint. The annular ligament attaches to the margin of the trochlear notch of the ulna, encircling the head of the radius. This ligament keeps the head of the radius in contact with the radial notch of the ulna.

Three major bursae are located at the elbow joint. The subcutaneous olecranon bursa is found in the connective tissue over the olecranon.

The intratendinous olecranon bursa is found in the triceps brachii tendon. The subtendinous olecranon bursa reduces friction between the triceps tendon and the olecranon, proximal to its insertion on the olecranon.

▶ **THE ROLE OF THE LIGAMENTS**

The ligaments of the elbow keep the joint secure, preventing medial and lateral deviation and keeping the radius in contact with the ulna.

LIGAMENTS OF THE ELBOW

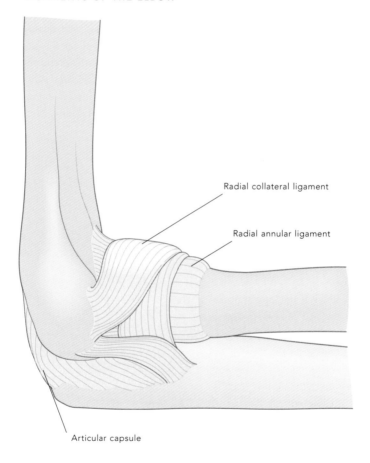

Radial collateral ligament

Radial annular ligament

Articular capsule

Muscles of the elbow

Anterior view of the elbow

THE BRACHIALIS IS THE MAIN FLEXOR OF THE ELBOW THAT ORIGINATES ON THE DISTAL HALF OF THE HUMERUS AND INSERTS ON THE TUBEROSITY OF THE ULNA. This muscle has the largest physiological cross-sectional area of the elbow flexors, and its ability to flex the elbow is not compromised by the position of the shoulder joint or the radioulnar joint.

The biceps brachii is activated maximally when it is flexing the elbow and supinating the forearm simultaneously. It can exert less force when the forearm is held in pronation or when the shoulder is fully flexed.

Brachioradialis is the longest of the elbow muscles and arises from the outer edge of the lower third of the humerus, crossing the elbow joint and inserting at the lower end of the radius near the styloid process. This muscle will flex the elbow and aid both pronation and supination—it will pronate the forearm from a supinated position and supinate the forearm from a pronated position.

Pronator teres pronates the forearm and contributes to elbow flexion. Its origins are on the medial epicondyle of the humerus and the coronoid process of the ulna, and it inserts on the lateral surface of the mid-radius.

▶ **ELBOW MOVEMENT**
The muscles of the upper arm produce movement at the elbow. The triceps brachii initiates elbow extension, while the biceps brachii initiates elbow flexion and also supinates the forearm.

ELBOW MUSCLES

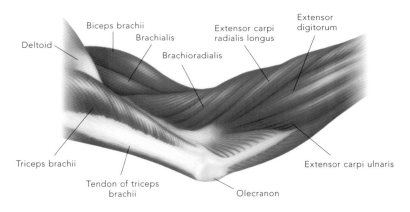

Deltoid

Biceps brachii

Brachialis

Brachioradialis

Extensor carpi
radialis longus

Extensor
digitorum

Triceps brachii

Tendon of triceps
brachii

Olecranon

Extensor carpi ulnaris

ELBOW JOINT—BACK

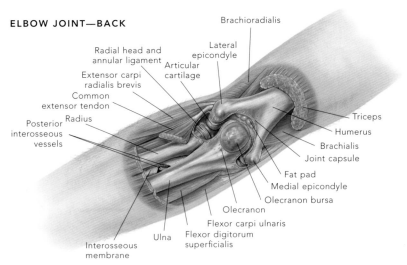

Brachioradialis

Radial head and
annular ligament

Lateral
epicondyle

Extensor carpi
radialis brevis

Articular
cartilage

Common
extensor tendon

Posterior
interosseous
vessels

Radius

Triceps

Humerus

Brachialis

Joint capsule

Fat pad

Medial epicondyle

Olecranon bursa

Olecranon

Flexor carpi ulnaris

Ulna

Flexor digitorum
superficialis

Interosseous
membrane

Posterior view of the elbow

THE TRICEPS BRACHII IS THE MAIN EXTENSOR OF THE ELBOW AND MAKES UP THE MAJORITY OF THE MUSCLE MASS ON THE POSTERIOR ARM. The long head of the triceps originates on the infraglenoid tubercle, allowing it to contribute to shoulder extension. The long head has the largest volume of all the muscles acting on the elbow joint. The medial and lateral heads only act to extend the elbow, with some of the fibers of the deeper medial head attaching to the joint capsule and drawing it taut during extension.

The anconeus is a small triangular muscle that assists the triceps in elbow extension. It arises from the lateral epicondyle of the humerus and inserts on the olecranon process and superior aspect of the elbow. While it is too small to make a great contribution to extension, it plays an important role in stabilizing the elbow joint during extension, pronation, and supination.

The supinator has two heads with origins on the lateral epicondyle of the humerus and the supinator fossa and crest of the ulna. It inserts on the lateral, posterior, and anterior surfaces of the proximal radius. Its function is to supinate the forearm by pulling on the radius at the radioulnar joint.

▶ **ELBOW MOVEMENT**
Contraction of the triceps brachii contributes to elbow extension, while the brachialis and brachioradialis contribute to elbow flexion.

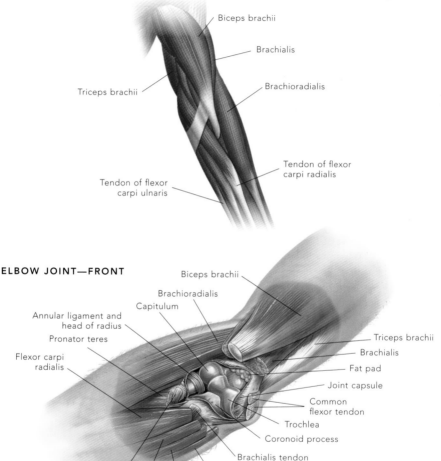

ELBOW MUSCLES—FRONT

Biceps brachii

Brachialis

Brachioradialis

Triceps brachii

Tendon of flexor
carpi radialis

Tendon of flexor
carpi ulnaris

ELBOW JOINT—FRONT

Biceps brachii

Brachioradialis

Capitulum

Annular ligament and
head of radius

Pronator teres

Flexor carpi
radialis

Triceps brachii

Brachialis

Fat pad

Joint capsule

Common
flexor tendon

Trochlea

Coronoid process

Brachialis tendon

Flexor carpi ulnaris

Flexor digitorum
superficialis

Palmaris
longus

Biceps femoris
tendon

Bones of the wrist

THE WRIST HAS A NUMBER OF BONES THAT FORM SEPARATE JOINTS. Just proximally to the wrist joint, the ulnar notch of the radius and the ulnar head articulate to form the distal radioulnar joint. This is a pivot joint, like the proximal radioulnar joint, and allows pronation and supination. The ulnar notch of the radius slides over the head of the ulna during these movements.

The carpal bones are arranged in two rows, proximal and distal. Laterally to medially the proximal row of bones is composed of the scaphoid, lunate, triquetrum, and pisiform. The distal row includes the trapezium, trapezoid, capitate, and hamate bones.

The concave surfaces of the radius articulate with the convex surfaces of the scaphoid and lunate bones to form the radiocarpal, or wrist, joint. The triquetrum also forms part of this joint when the wrist is completely adducted. The ulna does not form part of the wrist joint; it is separated from the carpal bones by an articular disc.

The eight carpal bones together are known as the carpus. The joint between the two rows of carpal bones is called the midcarpal joint. It is a series of synovial joints, and at the lateral portion of the joint, the scaphoid articulates with the trapezium and trapezoid. In the middle of the joint, the scaphoid and lunate articulate with the capitate, with the lunate also articulating with the hamate. The hamate also articulates with the triquetrum of the proximal row. Joints between each carpal bone and its adjacent bones are called the intercarpal joints.

▶ **CARPAL BONES**

The eight carpal bones—the trapezium, trapezoid, scaphoid, lunate, triquetral, pisiform, hamate, and capitate— connect the bones of the forearm to the bones of the hand.

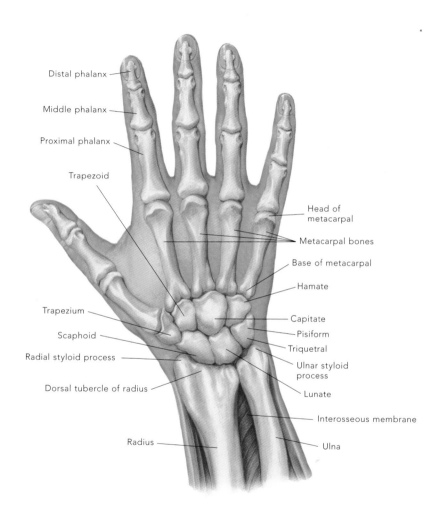

Distal phalanx

Middle phalanx

Proximal phalanx

Trapezoid

Head of metacarpal

Metacarpal bones

Base of metacarpal

Hamate

Trapezium

Capitate

Scaphoid

Pisiform

Radial styloid process

Triquetral

Ulnar styloid process

Dorsal tubercle of radius

Lunate

Interosseous membrane

Radius

Ulna

Ligaments and tendons of the wrist

THE WRIST JOINTS MAINTAIN STABILITY WITHIN A NUMBER OF MOBILE JOINTS WITH A COMPLEX CONFIGURATION OF LIGAMENTS. These consist of extrinsic ligaments that bridge from the carpal bones to the radius, and intrinsic ligaments that originate and insert on carpal bones.

The radiocarpal joint capsule has three thickenings to form capsular ligaments. The dorsal radiocarpal ligament extends from the posterior radius to the posterior scaphoid, lunate, and triquetrum. The palmar ulnocarpal ligament extends from the anterior edge of the articular disc and the styloid process of the ulna to the anterior surface of the carpal bones. The palmar radiocarpal ligament is a broad band of fibers of the capsule that pass from the anterior radius to the anterior surface of the proximal row of carpal bones.

The radiocarpal joint also has two collateral ligaments to limit adduction and abduction of the wrist. The radial collateral carpal ligament runs on the lateral side of the joint, from the styloid process of the radius to the scaphoid and the lateral side of the trapezium. The ulnar collateral carpal ligament runs from the ulnar styloid process to the pisiform and triquetrum. It is also continuous with the flexor retinaculum.

Joined by short bands of ligaments, the proximal carpals are loosely joined, while the distal carpals are more tightly bound to provide stability for metacarpals. Between the proximal and distal carpal rows, a number of intercarpal and collateral ligaments stabilize the midcarpal joint.

▶ **TENDON SHEATHS**

The tendons of the forearm muscles cross over the wrist area through lubricated tendon sheaths, which facilitate smooth movement.

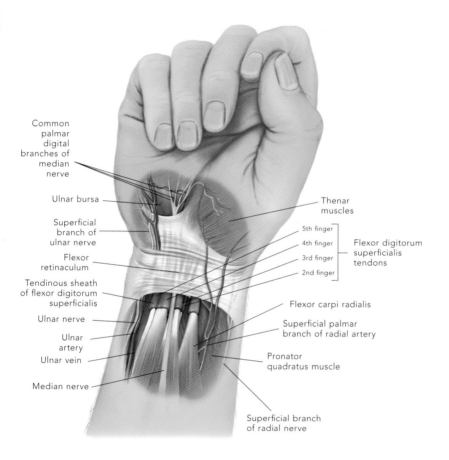

Common palmar digital branches of median nerve

Ulnar bursa

Superficial branch of ulnar nerve

Flexor retinaculum

Tendinous sheath of flexor digitorum superficialis

Ulnar nerve

Ulnar artery

Ulnar vein

Median nerve

Thenar muscles

5th finger
4th finger
3rd finger
2nd finger

Flexor digitorum superficialis tendons

Flexor carpi radialis

Superficial palmar branch of radial artery

Pronator quadratus muscle

Superficial branch of radial nerve

Muscles of the wrist

Anterior view of the wrist

THE ANTERIOR COMPARTMENT OF THE
FOREARM CONTAINS FLEXOR MUSCLES
THAT ACT ON BOTH THE WRIST AND
THE FINGERS.

Pronator quadratus is a deep muscle
on the distal forearm that acts to
pronate the forearm. It arises from the
anterior ulna and inserts on the
anterior surface of the radius. It works
in conjunction with pronator teres to
pronate the forearm and assists to hold
the distal radius and ulna together
when upward pressure is applied.

The superficial wrist flexor muscles
share a common origin at the medial
epicondyle of the humerus. The flexor
carpi ulnaris also originates from the
proximal ulna and runs down the
medial forearm, attaching to the
pisiform, hamate, and fifth metacarpal.
With the flexor carpi radialis, it flexes
the wrist, but when contracting with
the extensor carpi ulnaris, it adducts
the wrist. Flexor carpi radialis inserts
on the second and third metacarpals
and acts as a flexor of the wrist, but it
can also contribute to abduction,

pronation, and elbow flexion. Palmaris
longus is a smaller, weaker flexor
muscle that is absent in around 10% of
the population. It runs centrally
between the other superficial flexors
and attaches on the flexor retinaculum.

The flexor retinaculum in the wrist
consists of a group of strong connective
fibers that connect the lateral margin of
the radius, scaphoid, and trapezium
with the medial border of the ulna,
pisiform, and hamate. The branches
of the retinaculum to the underlying
bones create a series of sheath-like
compartments through which the
tendons of the flexor muscles of the
wrist and fingers pass.

▶ **HAND MOVEMENT**

A combination of tightly and loosely
bound carpal bones provide support to
the wrist, at the same time giving a
wide range of movements.

SURFACE MUSCLES OF THE WRIST

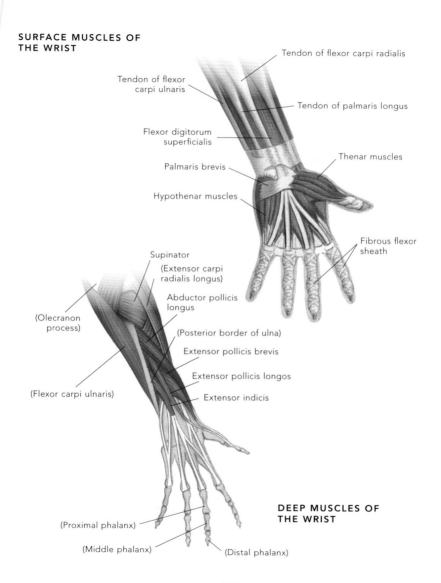

Tendon of flexor carpi radialis

Tendon of flexor carpi ulnaris

Tendon of palmaris longus

Flexor digitorum superficialis

Thenar muscles

Palmaris brevis

Hypothenar muscles

Fibrous flexor sheath

Supinator

(Extensor carpi radialis longus)

Abductor pollicis longus

(Olecranon process)

(Posterior border of ulna)

Extensor pollicis brevis

Extensor pollicis longos

(Flexor carpi ulnaris)

Extensor indicis

DEEP MUSCLES OF THE WRIST

(Proximal phalanx)

(Middle phalanx)

(Distal phalanx)

Posterior view of the wrist

THE POSTERIOR COMPARTMENT OF THE FOREARM IS COMPOSED OF MUSCLES THAT EXTEND THE WRIST AND FINGERS, many of which arise from the common extensor origin on the lateral epicondyle of the humerus.

The extensor carpi radialis longus muscle runs along the lateral side of the forearm, running from the lateral supracondylar ridge of the humerus to the base of the second metacarpal in the hand. It functions to extend the wrist and assists in abduction of the hand. The extensor carpi radialis brevis is a companion of the extensor carpi radialis longus, lying adjacent and partially covered by it. This muscle runs from the common extensor origin to the base of the third metacarpal and functions to extend the wrist. It also assists in abducting the hand. Extensor carpi ulnaris also originates at the common extensor origin and inserts on the medial base of the fifth metacarpal. Working with the other extensor muscles, it extends the wrist and assists with adduction of the wrist when working with the flexor carpi ulnaris.

The extensor retinaculum of the hand runs from the lateral margin of the radius to the medial carpal bones: the triquetrum and pisiform. Tendons of the extensor muscles of the wrist and fingers pass underneath the extensor retinaculum. As with the flexor retinaculum, it holds the tendons in place and prevents their movement away from the wrist during muscle contraction.

▶ **THE WRIST**
The tendons to the fingers and thumb, which enable movement, pass over the front and back of the wrist.

SURFACE MUSCLES OF THE WRIST

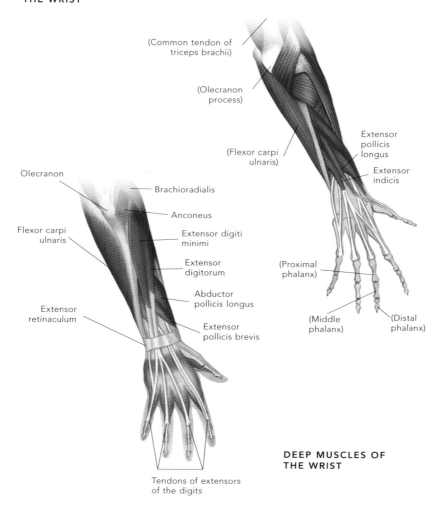

(Common tendon of triceps brachii)

(Olecranon process)

(Flexor carpi ulnaris)

Extensor pollicis longus

Extensor indicis

Olecranon

Brachioradialis

Anconeus

Extensor digiti minimi

Extensor digitorum

Abductor pollicis longus

(Proximal phalanx)

Flexor carpi ulnaris

Extensor retinaculum

Extensor pollicis brevis

(Middle phalanx)

(Distal phalanx)

Tendons of extensors of the digits

DEEP MUSCLES OF THE WRIST

Bones of the hand

DISTAL TO THE CARPAL BONES ARE THE BONES OF THE HAND, WHICH ARE KNOWN AS THE METACARPUS. There are five metacarpal bones that correspond to each digit. They are numbered in sequence from the lateral side; the most lateral on the hand proximal to the thumb is the first metacarpal, and the most medial is the fifth. Each metacarpal has a base, a body or shaft, and a head. The base of each metacarpal varies in shape to fit the articular surfaces of the carpals to form the common carpometacarpal joint. The bases of the second to fifth metacarpals also articulate with the adjacent metacarpals. Each body is slightly curved posteriorly. Each head is smooth and rounded to articulate with the concave base of the corresponding proximal phalanx, forming the five metacarpophalangeal joints.

There are 14 phalanges on each hand, three for each finger and two for the thumb. On the fingers the phalanges are referred to as the proximal, middle, and distal phalanges, while the thumb only has a proximal and a middle phalanx. Either side of the shaft of each phalanx is a proximal and a distal end, with the proximal end larger than the distal end or head. Their articulations form the interphalangeal joints and each phalanx is similar in shape and composition; however, the distal phalanges have relatively larger bases and heads in proportion to their length, where the head supports the pad of the fingertips.

▶ **HAND MOVEMENT**

The large number of bones in the hand, and the articulations between these bones, produce a highly mobile unit, capable of fine motor skills.

BONES OF THE HAND

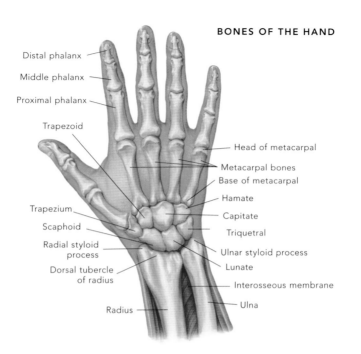

Distal phalanx

Middle phalanx

Proximal phalanx

Trapezoid

Head of metacarpal

Metacarpal bones

Base of metacarpal

Hamate

Trapezium

Capitate

Scaphoid

Triquetral

Radial styloid process

Ulnar styloid process

Dorsal tubercle of radius

Lunate

Interosseous membrane

Radius

Ulna

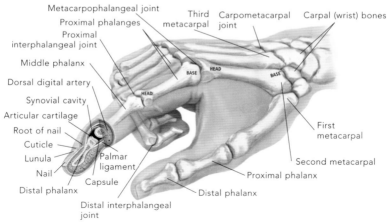

Metacarpophalangeal joint

Third metacarpal

Carpometacarpal joint

Carpal (wrist) bones

Proximal phalanges

Proximal interphalangeal joint

Middle phalanx

Dorsal digital artery

Synovial cavity

Articular cartilage

Root of nail

Cuticle

Lunula

Nail

Palmar ligament

Distal phalanx

Capsule

BASE

HEAD

HEAD

BASE

HEAD

First metacarpal

Second metacarpal

Proximal phalanx

Distal phalanx

Distal interphalangeal joint

Ligaments and tendons of the hand

THE CARPOMETACARPAL JOINTS ARE FORMED BY THE BASES OF THE MEDIAL FOUR METACARPALS WITH THE DISTAL ROW OF CARPAL BONES. As with all synovial joints, a fibrous capsule surrounds the common joint. On both the palmar (anterior) and dorsal (posterior) side, thickenings of the joint capsule form the capsular ligaments that support the joint. The dorsal carpometacarpal ligaments are a series of bands of fibers between the distal carpal bones to the bases of the metacarpals. Similarly, the palmar carpometacarpal ligaments pass between the carpal and metacarpal bones on the palmar aspect of the hand. At the base of the metacarpals are the intermetacarpal joints, which have the joint space that is continuous with the carpometacarpal joints and which are supported by the intermetacarpal ligaments.

The metacarpophalangeal joints are supported by a number of ligaments. The palmar ligament is a fibrocartilaginous plate that is firmly attached to the anterior base of the proximal phalanx and loosely attached to the neck of the metacarpal. It acts as an articular surface and facilitates flexion. The collateral ligaments pass from the side of the metacarpals, to the palmar aspect of the proximal phalanx and become taut during flexion. In addition, the heads of the second to fifth metacarpals are joined by deep transverse metacarpal ligaments, limiting their movement apart.

The proximal and distal interphalangeal joints are also supported by palmar and collateral ligaments. The palmar ligament also attaches to the fibrous flexor sheath of the digit, which is the common synovial sheath for the flexor tendons of the flexor digitorum superficialis and the flexor digitorum profundus.

▶ **MOVEMENT OF THE FINGERS**
There are no muscles in the fingers; movements of the finger tendons are initiated by the muscles of the palm and forearm.

LIGAMENTS AND TENDONS OF THE HAND

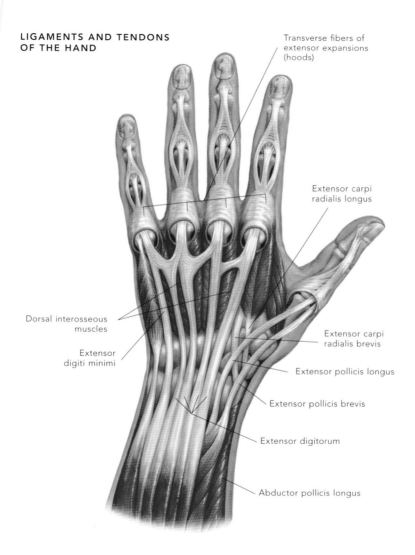

Transverse fibers of extensor expansions (hoods)

Extensor carpi radialis longus

Extensor carpi radialis brevis

Extensor pollicis longus

Extensor pollicis brevis

Extensor digitorum

Abductor pollicis longus

Dorsal interosseous muscles

Extensor digiti minimi

Muscles of the hand

Extensor muscles of the hand

ON THE DORSAL SIDE OF THE HAND ARE A NUMBER OF MUSCLES THAT EXTEND, ADDUCT, AND ABDUCT THE FINGERS. The muscles of the hand can be divided into two groups: the extrinsic muscles that originate in the forearm and the intrinsic muscles that are located within the hand itself.

Extensor digitorum originates at the common extensor origin and inserts on the dorsal digital expansion on the second to fifth digits. The dorsal digital expansion is a triangular aponeurosis that starts at the metacarpophalangeal joint, with fibers wrapping around to fuse together at the base of the distal phalanx. The extensor digitorum extends the metacarpophalangeal joint and assists in extension of the interphalangeal joints. Extensor digiti minimi also arises from the common extensor origin and inserts onto the dorsal digital expansion of the fifth digit. It extends the metacarpophalangeal joint and assists in extension of the interphalangeal joints of that digit. The last extrinsic

extensor is the extensor indicis, which lies deep to the extensor digitorum. It originates on the posterior surface of the ulna and inserts on the dorsal digital expansion. This muscle extends the second metacarpophalangeal joint and allows it to be used independently.

The dorsal interossei are intrinsic muscles that lie in the spaces between the metacarpals arising from the sides of the adjacent metacarpal bones. Inserting onto the proximal phalanx and the dorsal digital expansion, they abduct the index, middle, and ring fingers while also producing flexion of the metacarpophalangeal joint and

▶ **MOVEMENT OF THE HAND**

Though the dexterity of the hand is controlled in large part by muscles in the forearm, muscles in the hand play an important role in the sophisticated motor abilities of the hand.

extension of the interphalangeal joints. The lumbricals are four muscles that arise from the flexor tendons and insert on the lateral edge of the dorsal digital expansion. Their action is complex, as they help coordinate movements of both flexion and extension in fine movements such as writing. The lumbrical action involves flexion of the metacarpophalangeal joint and extension of the interphalangeal joints.

MUSCLES OF THE HAND

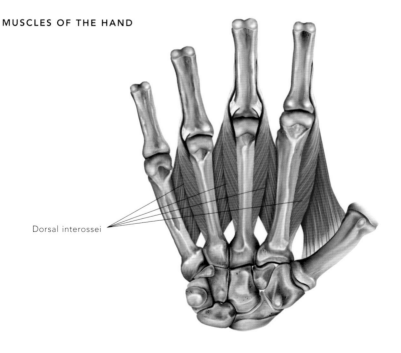

Dorsal interossei

Flexor muscles of the hand

AS WITH THE EXTENSOR MUSCULATURE, THE FLEXOR MUSCLES OF THE HAND CAN BE DIVIDED INTO TWO GROUPS: THE EXTRINSIC AND THE INTRINSIC MUSCLES.

Flexor digitorum superficialis lies deep to the wrist flexor muscles, arising from the common flexor origin—the coronoid process of the ulna and the upper anterior radius. It passes beneath the flexor retinaculum and inserts on the base of the second to fifth middle phalanges. This muscle flexes the metacarpophalangeal and proximal interphalangeal joints. Flexor digitorum profundus is also an extrinsic muscle originating on the medial and anterior surface of the proximal humerus. Inserting on the base of the second to fifth distal phalanges, this muscle primarily flexes the distal proximal interphalangeal joints and assists in flexion of the proximal interphalangeal and metacarpophalangeal joints.

As stated in the previous section, the lumbricals, with their unique attachments, will flex the metacarpophalangeal joints. There is one other intrinsic flexor of the fingers: the flexor digiti minimi brevis. Its origin is on the hamate and the flexor retinaculum, inserting on the base of the fifth proximal phalanx. This muscle flexes the metacarpophalangeal joint of the little finger.

The thumb has a number of muscles of its own that allow it to move more freely and exert more force than the fingers. It has two flexors: the flexor pollicis longus, an extrinsic muscle, and the flexor pollicis brevis, an intrinsic muscle. It has two extensors, the extensor pollicis longus and extensor pollicis brevis, which are both extrinsic muscles. It also has muscles that bring the thumb away from the

▶ HAND MOVEMENTS

The muscles of the forearm and hand combine to produce dexterity in the hands and fingers.

hand or toward the hand, and a movement called opposition, where the thumb is brought forward and toward the other fingers so that it can come into precise contact with any finger.

These muscles are the abductor pollicis longus, abductor pollicis brevis, opponens pollicis, adductor pollicis, and palmaris brevis.

FLEXOR MUSCLES OF THE HAND

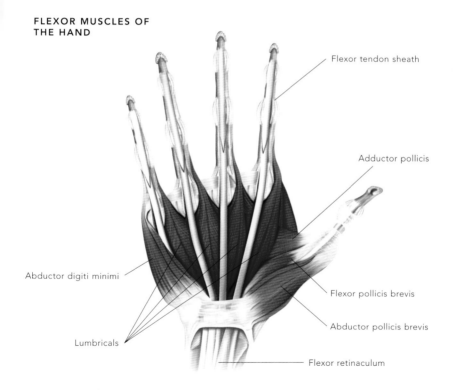

Flexor tendon sheath

Adductor pollicis

Abductor digiti minimi

Flexor pollicis brevis

Abductor pollicis brevis

Lumbricals

Flexor retinaculum

The movement of throwing

Throwing is a complex movement that is common to a number of sports. It requires coordination through the whole body to transfer energy to the ball and to control the precise speed and trajectory.

The early stages of the throw involve the transfer of weight to the rear foot and rotation of the torso to draw the throwing arm away from the target. In order to generate power in the throw, the muscles involved must be stretched. The pectoral girdle is retracted and depressed by the trapezius, stretching the serratus anterior and pectoralis minor. The shoulder joint is pulled into extension by the latissimus dorsi and teres major, stretching the pectoralis major and the anterior deltoid fibers. The mid-deltoid muscles pull the arm into abduction. The arm is laterally rotated by the posterior deltoid fibers and infraspinatus.

As the throw moves through the cocking stages, the weight of the body is shifted forward by the lower limbs and through rotation of the trunk. The pectoral girdle is rapidly protracted by the serratus anterior and pectoralis minor. Anterior deltoid fibers and the pectoralis major flex the shoulder joint, and the subscapularis, having been fully stretched, medially rotates the arm.

▶ **THROWING MOTION**

The throwing motion in five stages: wind up (A); stride (B); late cocking (C); acceleration (D); follow-through (E).

LATE COCKING PHASE

THE THROWING MOTION

A   B   C   D   E

The movement of throwing (cont.)

Late in the cocking phase, the arm is in a position of extreme external rotation. This helps the thrower generate speed; however, it also forces the head of the humerus forward, which places significant stress on the capsular ligaments in the front of the shoulder, the glenoid labrum, and the tendons of muscles such as the biceps and supraspinatus. Over time, the ligaments become more lax, resulting in greater external rotation range of motion and less shoulder stability.

Significant stress is also placed on the elbow in this position. The ulnar collateral ligament must resist high forces that are pulling the elbow joint into abduction. Stress is also placed on the tendon of the common flexor origin on the medial epicondyle of the humerus.

Once the ball is released, the ligaments and rotator cuff muscles at the back of the shoulder are exposed to high stresses to decelerate the arm and control the humeral head. Infraspinatus and teres minor contract eccentrically to slow the rapid internal rotation of the shoulder. The posterior fibers of the rear deltoid contract to limit extension on the follow-through of the throw. The posterior capsule of the shoulder joint is subjected to high tensile forces that can, over time, lead to hypertrophy. This causes the athlete to lose range of motion in internal rotation. This, combined with excess external rotation, results in a greater risk of labrum and rotator cuff tears.

A superior labrum anterior to posterior (SLAP) tear is a common throwing injury. This involves damage to the fibrocartilage rim, called the glenoid labrum, in the shoulder joint. With a SLAP injury, the humeral head acts as a lever and tears the biceps tendon and labrum cartilage from the glenoid bone in a front-to-back (anterior–posterior) direction.

TEAR IN THE LABRUM

▼ **THROWING INJURIES**

The anatomy is put under considerable stress during a throw. This sometimes results in a rupture in the tissues, such as a labrum tear.

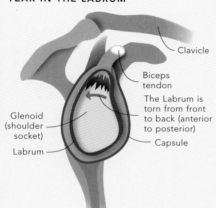

Clavicle

Biceps tendon

The Labrum is torn from front to back (anterior to posterior)

Glenoid (shoulder socket)

Labrum

Capsule

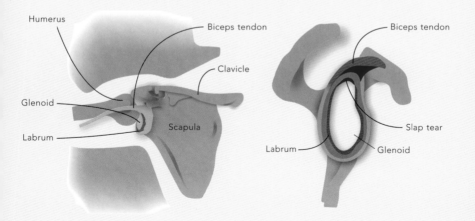

Humerus

Biceps tendon

Clavicle

Glenoid

Scapula

Labrum

Biceps tendon

Slap tear

Labrum

Glenoid

Chapter 10:
Pelvic girdle and lower limb

The pelvic girdle and lower limbs, or legs, play a crucial role in the body's preferred forms of locomotion, walking and running. They are capable of producing and withstanding large forces, and in some individuals, if required, they can help the body achieve speeds in excess of 20 miles per hour. They also act as a stable platform, so that the upper body and limbs can produce complex movements.

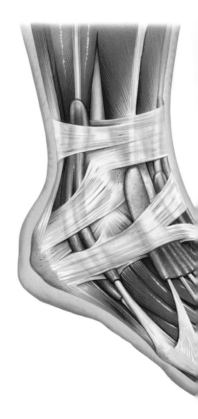

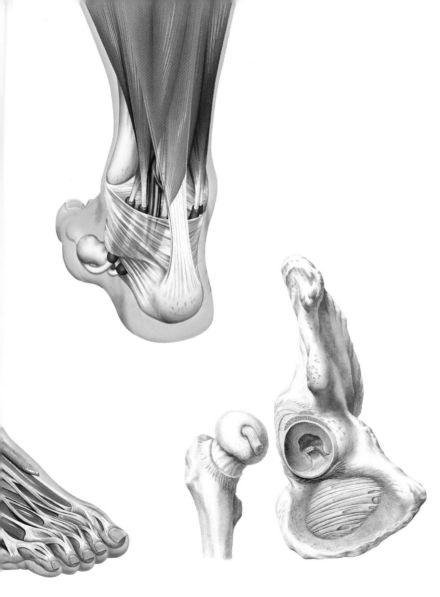

Bones of the lower limb

THE PELVIS CONSISTS OF A PAIR OF LARGE, FLAT BONES THAT JOIN THE LOWER LIMB APPENDICULAR SKELETON TO THE AXIAL SKELETON. The pelvic bones extend anteriorly and laterally from the sacrum at the sacroiliac joints. The left and right hip bones meet anteriorly at the body's midline in a band of fibrocartilage called the pubic symphysis. The hip bones also form the hip joint with the femur.

The only bone in the thigh, the femur, is the longest, heaviest, and strongest bone in the body. The femur needs to carry the weight of the body and withstand forces far in excess of body weight during activities such as walking and running. It also needs to accommodate large forces produced by the muscles that attach to it.

In anatomical terms, the leg is the body segment between the knee and the ankle. There are two bones in the leg, of which the tibia performs the vast majority of weight bearing, while the fibula provides a site for muscle attachments. The tibia is widest at its proximal end, where it forms the knee joint with the femur.

Distal to the tibia are the tarsal bones of the ankle and foot. The talus is the most proximal of these bones and forms the ankle joint with the tibia and fibula. It transmits forces from the tibia to the heel bone, known as the calcaneus. The metatarsals form the middle of the foot distal to the tarsals. As with the bones of the fingers, the bones that form the toes are called phalanges.

▶ **THE LOWER LIMB**

The arrangement of the bones of the lower limb is similar to those of the arm, with a large, strong bone in the upper area and two smaller bones in the lower region.

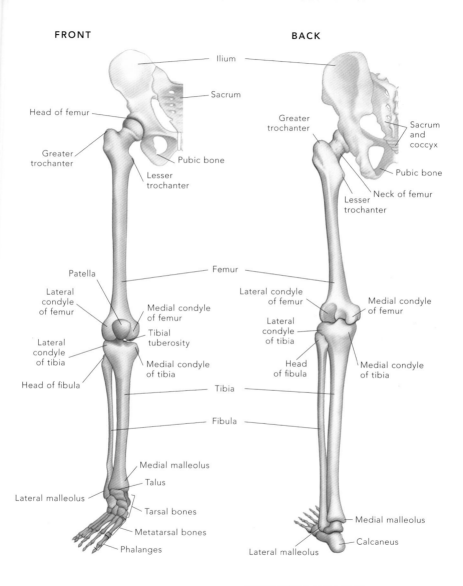

FRONT

BACK

Ilium

Sacrum

Head of femur

Greater trochanter

Pubic bone

Lesser trochanter

Greater trochanter

Sacrum and coccyx

Pubic bone

Neck of femur

Lesser trochanter

Patella

Femur

Lateral condyle of femur

Medial condyle of femur

Tibial tuberosity

Medial condyle of tibia

Lateral condyle of tibia

Head of fibula

Lateral condyle of femur

Medial condyle of femur

Lateral condyle of tibia

Head of fibula

Medial condyle of tibia

Tibia

Fibula

Medial malleolus

Talus

Lateral malleolus

Tarsal bones

Metatarsal bones

Phalanges

Medial malleolus

Calcaneus

Lateral malleolus

Muscles of the lower limb
Deep muscles of the lower limb

THE GLUTEAL REGION IS AN ANATOMICAL AREA LOCATED POSTERIORLY TO THE PELVIC GIRDLE, AT THE PROXIMAL END OF THE FEMUR. The smaller, deep muscles of the hip stabilize the joint and provide lateral rotation of the lower limb. They also stabilize the hip joint by pulling the femoral head into the acetabulum of the pelvis. This group includes the quadratus femoris, obturator internus, piriformis, gemellus superior, and gemellus inferior.

The muscles in the medial compartment of the thigh are known as the hip adductors. These muscles move the thigh toward the body's midline. Included in this group are the adductor longus, adductor brevis, adductor magnus, pectineus, and gracilis muscles.

The quadriceps muscles are found in the anterior compartment of the thigh. They are the main extensors of the knee joint. Of the four muscles that compose this group, the vastus intermedius lies deep to the others. The other quadriceps muscles are described in the following section on surface muscles of the lower limb.

The soleus is an important plantar flexor of the ankle joint. This muscles lies in the posterior compartment of the leg, deep to the gastrocnemius muscle. Other plantar flexor muscles found deep in this compartment include the plantaris and tibialis posterior.

▶ **THE QUADRICEPS**
The four parts of the quadriceps muscle taper down and merge to become a single tendon, which extends across the knee joint and attaches to the tibia.

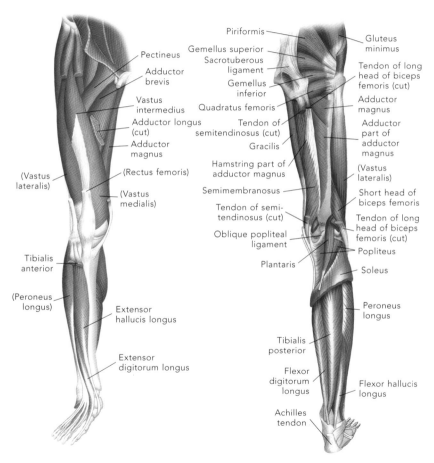

FRONT

BACK

Pectineus

Adductor
brevis

Vastus
intermedius

Adductor longus
(cut)

Adductor
magnus

(Vastus
lateralis)

(Rectus femoris)

(Vastus
medialis)

Tibialis
anterior

(Peroneus
longus)

Extensor
hallucis longus

Extensor
digitorum longus

Piriformis

Gemellus superior

Sacrotuberous
ligament

Gemellus
inferior

Quadratus femoris

Tendon of
semitendinosus (cut)

Gracilis

Hamstring part of
adductor magnus

Semimembranosus

Tendon of semi-
tendinosus (cut)

Oblique popliteal
ligament

Plantaris

Tibialis
posterior

Flexor
digitorum
longus

Achilles
tendon

Gluteus
minimus

Tendon of long
head of biceps
femoris (cut)

Adductor
magnus

Adductor
part of
adductor
magnus

(Vastus
lateralis)

Short head of
biceps femoris

Tendon of long
head of biceps
femoris (cut)

Popliteus

Soleus

Peroneus
longus

Flexor hallucis
longus

Surface muscles of the lower limb

THE GLUTEUS MAXIMUS IS THE LARGEST MUSCLE IN THE BODY. IT LIES SUPERFICIALLY, COVERING MOST OF THE OTHER GLUTEAL MUSCLES. The gluteus maximus muscle transmits its force to the femur to extend the hip joint and also attaches a wider surface composed of the iliotibial tract. The iliotibial tract is a longitudinal thickening of the fascia lata, located laterally in the thigh. The fascia lata is a deep fascial supportive structure of the whole thigh musculature. It begins proximally around the iliac crest and extends distally to the bony prominences of the tibia, where it continues to become the deep fascia of the leg.

On the thigh, the most superficial and prominent muscles in the anterior compartment are three of the four quadriceps muscles. The rectus femoris is located in the middle of the thigh, while the vastus medialis and vastus lateralis are located on the medial and lateral sides of the femur, respectively. In addition to the quadriceps, the sartorius is also located in the anterior compartment; it extends from the anterior superior iliac spine of the pelvis and runs across the anterior thigh to insert on the anteromedial surface of the upper tibia. Finally, the pectineus is the most anterior adductor of the hip that functions to adduct and medially rotate the thigh.

The muscles in the posterior compartment of the thigh are collectively known as the hamstrings. They consist of the biceps femoris, semitendinosus, and semimembranosus. The adductor muscles lie in the medial compartment.

The gastrocnemius is the most prominent and superficial muscle in the posterior leg. The shape of the calf is mainly due to the two fleshy bellies of this muscle. All other muscles in the posterior leg lie deep to the gastrocnemius, which performs plantar flexion of the foot as well as flexion of the knee.

There are four muscles in the anterior compartment of the leg, including tibialis anterior, extensor digitorum longus, extensor hallucis longus, and peroneus tertius. These muscles act as a group to dorsiflex the ankle and invert the foot.

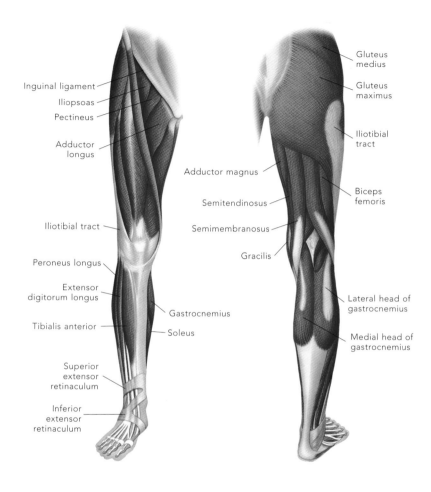

Inguinal ligament
Iliopsoas
Pectineus
Adductor longus
Iliotibial tract
Peroneus longus
Extensor digitorum longus
Tibialis anterior
Superior extensor retinaculum
Inferior extensor retinaculum

Gastrocnemius
Soleus

Gluteus medius
Gluteus maximus
Iliotibial tract
Adductor magnus
Biceps femoris
Semitendinosus
Semimembranosus
Gracilis
Lateral head of gastrocnemius
Medial head of gastrocnemius

▲ **THE LOWER LIMB**

The muscles of the lower limb are compartmentalized by sheets of connective tissue.

Bones of the hip joint

THE PELVIS IS MADE UP OF TWO SYMMETRICAL HIP BONES, WHICH ARE KNOWN AS THE INNOMINATE BONES OR PELVIC BONES. Each hip bone comprises three bones that are fused together: the ilium, the ischium, and the pubis.

The ilium is the widest and largest of the three bones and forms the superior part of the hip. Superior to the hip joint, the ilium forms a prominent broad blade for the attachment of ligaments and large muscles. The inner surface is concave and is known as the iliac fossa. The external surface is convex and provides attachments to the gluteal muscles, hence it is known as the gluteal surface.

The ischium is the posterior inferior part of the hip bone and is composed of a body and an inferior and a superior ramus (branches). The inferior ischial ramus combines with the inferior pubic ramus, forming the ischiopubic ramus. The posteroinferior aspect of the ischium forms the ischial tuberosity, which, in the sitting position, is where the weight of the body rests.

The pubis is an angulated bone that forms the most anterior portion of the hip bone. It consists of a body and superior and inferior rami. The body extends medially, articulating with the pubic body of the other side at the cartilaginous pubic symphysis joint.

The three hip bones join to form part of the acetabulum, the articular surface of the hip that forms the ball-and-socket hip joint with the head of the femur.

▶ **ACETABULUM**

The three hip bones that make up part of the acetabulum fuse during adolescence. The tuberosities of the ischium are the bony protuberances we sit on.

PELVIS AND HIP

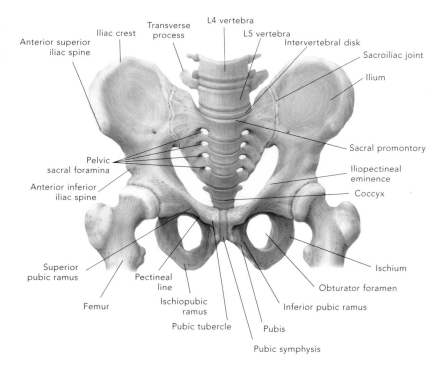

Anterior superior iliac spine

Iliac crest

Transverse process

L4 vertebra

L5 vertebra

Intervertebral disk

Sacroiliac joint

Ilium

Sacral promontory

Pelvic sacral foramina

Anterior inferior iliac spine

Iliopectineal eminence

Coccyx

Superior pubic ramus

Femur

Pectineal line

Ischiopubic ramus

Pubic tubercle

Pubis

Pubic symphysis

Inferior pubic ramus

Obturator foramen

Ischium

Ligaments and tendons of the hip joint

THE HIP JOINT HAS A STRONG, SUPPORTIVE NETWORK OF CONNECTIVE TISSUE TO PROVIDE STABILITY TO THE JOINT. This, in combination with the shape of the joint where the hemispherical head of the femur fits completely into the cup-shaped acetabulum, makes this joint very stable. There is a fibrocartilaginous rim around the acetabulum known as the acetabular labrum. This increases its depth and thus improves the stability of the joint.

The joint capsule is very strong, with fibers arranged in a number of directions to withstand tensile stresses. There are a number of identifiable capsular ligaments reinforcing the capsule. The iliofemoral ligament prevents hyperextension of the hip joint. It originates from the ilium and attaches to the femur between the trochanters, on the intertrochanteric line. The ischiofemoral ligament is a spiral-shaped ligament that originates from the ischium and attaches to the greater trochanter of the femur. It limits excessive extension at the hip joint. The pubofemoral ligament strengthens the inferior and anterior aspects of the capsule and prevents excessive abduction and extension. It originates at the iliopubic eminence and obturator membrane and then blends with the articular capsule, inserting in the intertrochanteric line. The intracapsular ligamentum teres is a weak band of connective tissue that runs from the acetabular fossa to the fovea on the femoral head.

▶ **THE LIGAMENTS**

The iliofemoral, ischiofemoral, and pubofemoral ligaments are classed as extracapsular ligaments. The ligamentum teres is classed as an intracapsular ligament.

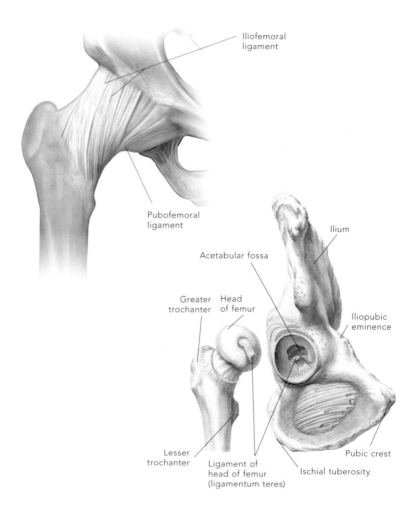

Iliofemoral
ligament

Pubofemoral
ligament

Acetabular fossa

Ilium

Greater
trochanter

Head
of femur

Iliopubic
eminence

Lesser
trochanter

Ligament of
head of femur
(ligamentum teres)

Ischial tuberosity

Pubic crest

Muscles of the hip joint
Muscles of the front of the hip joint

AT THE FRONT OF THE HIP LIE A NUMBER OF MUSCLES THAT FLEX THE HIP. The psoas major is a large, thick muscle that originates on the vertebral bodies and transverse processes of T12 to L5 and converges to insert on the lesser trochanter of the femur. It flexes the hip and can also contribute to flexion of the lumbar spine. The iliacus also inserts on the lesser trochanter, blending with psoas major. It originates at the iliac fossa of the ilium. The rectus femoris is one of the quadriceps muscles. It is the only quadriceps muscle that crosses the hip joint, allowing it to contribute to hip flexion as well as knee extension. It originates at the anterior inferior iliac spine and just above the acetabulum. The sartorius is the most superficial of the anterior leg muscles. It is a long strip of muscle that runs from the anterior superior iliac spine, wrapping around the front of the thigh to attach to the medial shaft of the tibia. It will produce many of the movements that combine to produce a cross-legged

sitting position, including hip flexion and abduction, and knee flexion. The pectineus is a rectangular-shaped muscle located deep in the groin. It spans between the pubis and the lesser trochanter. The pectineus both flexes and adducts the hip.

The adductor magnus is the largest of the adductor muscles. It has an adductor component and a hamstring component. It originates at the ischiopubic ramus and inserts on the medial condyle of the femur and along the shaft of the femur in a fan shape. The adductor longus is a thin, triangular muscle lying over the top of the adductor magnus. It originates at the pubis, and its fibers fan out to attach along the shaft of the femur anteriorly to the adductor mangnus. The adductor brevis has a similar shape and origin, with its fibers fanning out to attach along the upper half of the femoral shaft.

SURFACE MUSCLES

DEEP MUSCLES

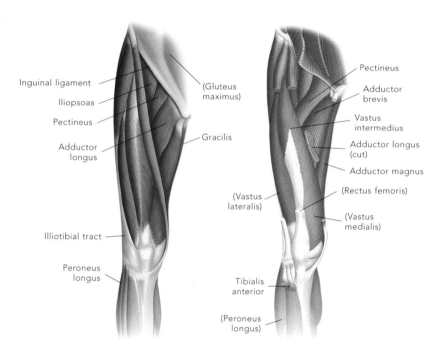

Inguinal ligament

(Gluteus maximus)

Iliopsoas

Pectineus

Gracilis

Adductor longus

Illiotibial tract

Peroneus longus

Pectineus

Adductor brevis

Vastus intermedius

Adductor longus (cut)

Adductor magnus

(Rectus femoris)

(Vastus lateralis)

(Vastus medialis)

Tibialis anterior

(Peroneus longus)

▲ HIP MUSCLES

The hip region contains powerful muscles, including the largest muscle in the body, the gluteus maximus.

Muscles of the rear of the hip joint

THE HIP EXTENSORS ARE FOUND AT THE BACK OF THE HIP AND THE MAIN HIP EXTENSOR—and largest of the gluteal muscles—is the gluteus maximus. It originates from the gluteal surface of the ilium, sacrum, and coccyx and inserts into the iliotibial tract and greater trochanter of the femur. The gluteus maximus extends the hip and contributes to abduction and lateral rotation.

The hamstring muscle group is composed of the semimembranosus, semitendinosus, and biceps femoris. The hamstring muscles are the main flexors of the knee joint; however, they also originate on the ischial tuberosity and cross the hip joint, so they can simultaneously produce hip extension when contracting.

There are a number of gluteal muscles that are deep to the gluteus maximus and abduct the thigh at the hip joint. The gluteus medius is only partially covered by the gluteus maximus; it has its origin on the gluteal surface of the ilium and converges to a flattened tendon inserting on the greater trochanter.

The gluteus minimus is the smallest of the gluteal muscles, and it runs from the gluteal surface—anteriorly and inferiorly to the gluteus medius—and inserts on the greater trochanter.

In this region of the hip are the lateral rotators of the hip. While the gluteus maximus also performs this function, the piriformis, obturator internus, gemellus superior, gemellus inferior, quadratus femoris, and obturator externus primarily generate this movement. All insert on the greater trochanter, and all arise from the ischium except for the piriformis, which originates on the sacrum.

▶ **THE GLUTEAL MUSCLES**
The gluteus maximus extends the thigh, while the gluteus medius and gluteus minimus work to keep the pelvis level and swing the opposite side forward during walking.

MUSCLES OF THE REAR OF THE HIP JOINT

Gluteus medius

Gluteus maximus

Iliotibal tract

Adductor magnus

Semitendinosus

Biceps femoris

Semimembranosus

Gracilis

Piriformis

Gemellus superior

Sacrotuberous ligament

Gemellus inferior

Quadratus femoris

Tendon of semitendinosus (cut)

Gracilis

Hamstring part of adductor magnus

Semimembranosus

Tendon of semi-tendinosus (cut)

Oblique popliteal ligament

Gluteus minimus

Tendon of long head of biceps femoris (cut)

Adductor magnus

Adductor part of adductor magnus

(Vastus lateralis)

Short head of biceps femoris

Tendon of long head of biceps femoris (cut)

Bones of the knee joint

THE KNEE JOINT IS A HINGE JOINT FORMED BY ARTICULATIONS BETWEEN THE PATELLA, FEMUR, AND TIBIA. The tibiofemoral articulation is formed by the condyles of the femur and the proximal tibia. The patellofemoral joint is formed by the articulation of the anterior distal femur and the patella.

The distal femur is characterized by the rounded medial and lateral condyles. The two convex condyles articulate with the tibia. Hyaline cartilage covers both condyles and the patellar surface, which unites the two. The intercondylar fossa is a depression found on the posterior surface of the femur between the two condyles and contains two facets for attachment of internal knee ligaments.

The proximal tibia is widened by the medial and lateral condyles, which assist in weight bearing. On the upper surface of each tibial condyle is a flat articular surface covered in hyaline cartilage and separated by the intercondylar eminence. This is the main site of attachment for the ligaments and the menisci of the knee.

The patella is a sesamoid bone located at the front of the knee joint within the patellofemoral groove of the femur. It attaches superiorly to the quadriceps tendon and inferiorly to the patellar ligament. The articular surface of the patella is oval-shaped and is covered in extremely thick cartilage due to the very high forces it is subjected to during gait.

▶ **BONES OF THE KNEE**
The knee joint is a complex joint that plays a major role in weight bearing and mobility.

FEMUR

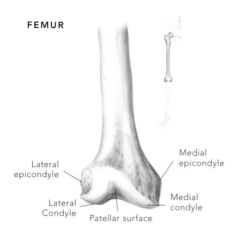

Lateral
epicondyle

Medial
epicondyle

Lateral
Condyle

Medial
condyle

Patellar surface

FIBULA

Apex of fibula

Neck of fibula

Lateral surface

Anterior border

Medial surface

Head of fibula

TIBIA

Intercondylar eminence

Lateral condyle

Superior articular
surfaces (medial
and lateral facets)

Articular surface
with head of fibula

Medial
condyle

Oblique line

Tibial tuberosity

Anterior
intercondylar area

Lateral surface

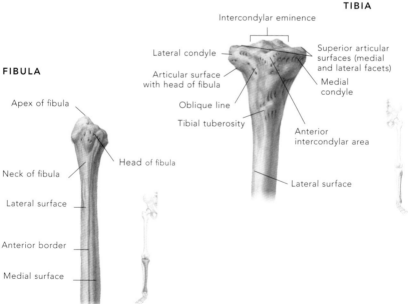

Ligaments and tendons of the knee joint

THE KNEE JOINT IS SUPPORTED BY THE MENISCI, WHICH ARE TWO CRESCENT-SHAPED FIBROCARTILAGE STRUCTURES THAT SIT ON BOTH ENDS OF THE INTERCONDYLAR AREA OF THE TIBIA. They act to deepen the articular surface of the tibial condyles and assist in weight bearing across the joint, aid lubrication, and act as shock absorbers.

The knee has two intracapsular ligaments, the anterior and posterior cruciate ligaments, which cross each other between their attachments. The posterior cruciate ligament limits the posterior translation—or sliding of the femur off the tibia—and the anterior displacement of the femur on a fixed tibia. It attaches at the posterior intercondylar region of the tibia and ascends anteriorly to attach to the femur in the intercondylar fossa. The anterior cruciate ligament limits the anterior translation of the tibia. It attaches at the anterior intercondylar region of the tibia and ascends posteriorly to attach to the femur in the intercondylar fossa.

The joint capsule of the knee also strongly supports the joints and is composed of muscle tendons and their expansions—areas of flat connective tissue that arise from the tendon or ligament. The oblique popliteal ligament is an expansion of the semimembranosus tendon. The arcuate popliteal ligament arises from the fused tendons of the quadriceps muscles.

Two extracapsular collateral ligaments act to prevent any medial or lateral movement. The tibial (or medial) collateral ligament is found on the medial side of the joint. Proximally, it attaches to the medial epicondyle of the femur, while distally, it attaches to the medial surface of the tibia. The fibular (or lateral) collateral ligament is a thinner band of fibers that attaches proximally to the lateral epicondyle of the femur and distally attaches to a depression on the lateral surface of the fibular head.

The ligamentum patellae is distal to the patella and is a continuation of the quadriceps tendon. It is a strong, flat band of fibers that attach to the tibial tuberosity.

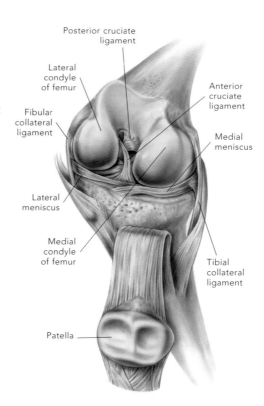

Posterior cruciate ligament

Lateral condyle of femur

Anterior cruciate ligament

Fibular collateral ligament

Medial meniscus

Lateral meniscus

Medial condyle of femur

Tibial collateral ligament

Patella

▶ **LIGAMENT STABILITY**

The ligaments are thickest on the sides of the joint and are arranged to limit lateral movement, rotation, and sliding of the femur on the tibia. The hyaline cartilage allows for smooth flexion and extension movements.

Muscles of the knee joint
Muscles of the front of the knee joint

IN THE ANTERIOR COMPARTMENT OF THE THIGH IS THE QUADRICEPS FEMORIS, WHICH EXTENDS THE KNEE. This group of muscles is composed of four parts. The rectus femoris is a bipennate muscle that originates on the anterior inferior iliac spine and above the acetabulum. Two-thirds of the way down the thigh it narrows to a thick tendon attaching to the upper border of the patella. The vastus lateralis sits laterally to the rectus femoris; it arises from the intertrochanteric line and has a linear attachment along the proximal half of the femoral shaft. It inserts into both the tendon of the rectus femoris and the lateral border of the patella. On the medial aspect of the thigh is the vastus medialis, which originates at the lower intertrochanteric line and wraps around the back of the femur to a linear attachment on the medial shaft of the femur. It inserts onto the tendon of the rectus femoris, the medial border of the patella, and the medial condyle of the tibia.

Between the vastus lateralis and vastus medialis, and deep to the rectus femoris, is the vastus intermedius. It arises from the upper two-thirds of the anterior femur and inserts on the base of the patella. The fibers of the vastus intermedius blend extensively with the vastus lateralis and vastus medialis.

The quadriceps muscle group is the main extensor of the knee, with each muscle playing a slightly different role during different phases of extension. The rectus femoris is particularly active when a combined movement of flexing the hip and extending the knee is required—such as when kicking a ball.

▶ **MOVEMENT OF THE KNEE JOINT**

The strong muscles of the upper leg are primarily responsible for initiating knee joint movement.

SURFACE MUSCLES

DEEP MUSCLES

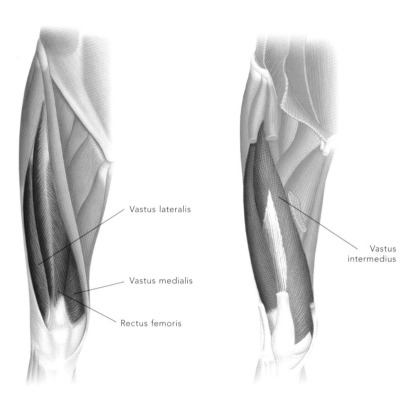

Vastus lateralis

Vastus medialis

Rectus femoris

Vastus intermedius

Muscles of the rear of the knee joint

THE HAMSTRING MUSCLES HAVE A COMPLEX ACTION AND COLLECTIVELY ACT AS THE MAIN FLEXORS OF THE KNEE. The hamstrings consist of the semitendinosus, semimembranosus, and the long and short heads of the biceps femoris. Except for the short head of the biceps femoris, all the hamstring muscles cross the knee and hip joints. The semitendinosus arises from the lateral ischial tuberosity and attaches to the medial condyle of the tibia. The semimembranosus lies mostly underneath the semitendinosus, and its origin is on the upper facet of the ischial tuberosity while its insertion is on the posterior medial tibial condyle. The biceps femoris, the most lateral of the hamstrings, has two heads that originate a considerable distance from each other. The long head originates with the tendon of the semitendinosus, and the short head arises from the lower lateral shaft of the femur. The fibers of the long and short heads blend together at the tendinous junction and insert onto the head of the fibula, with some fibers attaching on the lateral tibial condyle.

The gastrocnemius is primarily a plantar flexor of the ankle; however, as both the medial and lateral heads cross the knee joint, it is also a strong flexor of the knee. It attaches to the medial and lateral condyles of the femur, with the heads gradually blending together and narrowing to the broad Achilles tendon.

The popliteus muscle arises from the lateral condyle of the femur, crossing the knee joint and sweeping down and medially before attaching to the posterior surface of the tibia. The popliteus is a weak flexor of the knee, and its main role is to laterally rotate the femur on the tibia, releasing the knee from its locked position to allow flexion.

▶ **THE HAMSTRINGS**
The group of three muscles known as the hamstrings work to straighten the hip joint and bend the knee joint.

THE REAR OF THE
KNEE JOINT

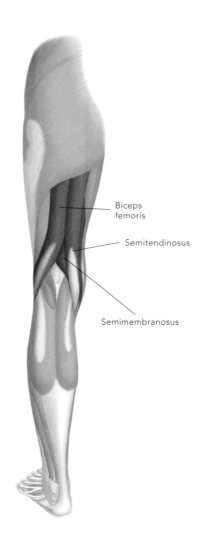

Biceps
femoris

Semitendinosus

Semimembranosus

Bones of the ankle

THE ANKLE JOINT IS A SYNOVIAL HINGE JOINT FORMED BY THE TIBIA, FIBULA, AND THE TALUS BONES OF THE FOOT. The ankle joint allows dorsiflexion and plantar flexion of the foot.

The distal tibia widens from its shaft to form the weight-bearing trochlear surface. A bony projection on the medial side, the medial malleolus, is also covered in cartilage to articulate with the tarsal bone. Laterally, the tibia has a fibular notch where it forms the inferior tibiofibular joint.

The distal end of the fibula forms the bony lateral prominence of the ankle: the lateral malleolus. The articulating surface of the tibia, rather than playing a weight-bearing role, is mostly a supporting structure for the joint. The socket formed by the distal tibia and fibula is known as the ankle mortise.

The talus is one of the metatarsal bones of the foot, and it forms the distal surface of the ankle joint articulating with the tibia and fibula. The body of the talus fits snugly into the ankle mortise. The articulating surface of the talus is wedge-shaped, and it is wider anteriorly and more narrow posteriorly. During plantar flexion and dorsiflexion, the talus is held in the ankle mortise and the joint is more stable.

▶ **BONES OF THE ANKLE**
The ankle is formed by the tibia, fibula, and talus.

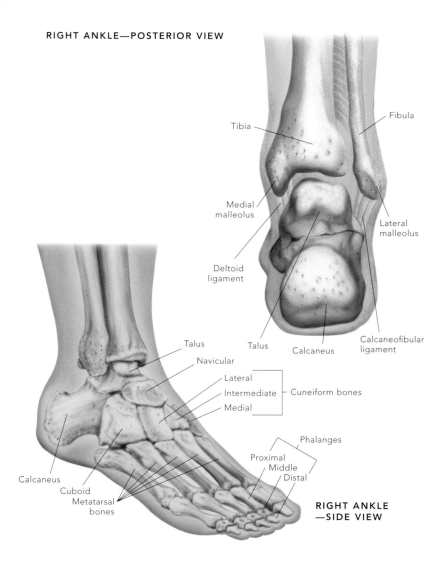

Tibia

Fibula

Medial
malleolus

Lateral
malleolus

Deltoid
ligament

Calcaneofibular
ligament

Talus

Calcaneus

Talus

Navicular

Lateral

Intermediate — Cuneiform bones

Medial

Phalanges

Proximal
Middle
Distal

Calcaneus

Cuboid

Metatarsal
bones

**RIGHT ANKLE
—SIDE VIEW**

Ligaments and tendons of the ankle

THE ANKLE JOINT IS A HINGE JOINT THAT IS SUPPORTED BY STRONG COLLATERAL LIGAMENTS THAT ALSO LIMIT LATERAL MOVEMENT. It is important to note that while the foot can invert and evert, this occurs at a joint lower in the foot, the subtalar joint.

The lateral collateral ligament limits adduction of the ankle joint and is composed of three distinct and separate ligaments. The anterior talofibular ligament limits anterior movement of the talus and spans from the lateral malleolus to the talus. The posterior talofibular ligament also runs from the lateral malleolus to the talus and limits posterior movement of the talus. The calcaneofibular ligament arises from the lateral malleolus, where it is fused with the other lateral collateral ligaments. This ligament attaches to the middle of the lateral calcaneus, or heel bone.

Medially, the deltoid ligament limits abduction of the ankle joint. Though only named as one ligament, the deltoid ligament has a number of parts, with fibers traveling in different directions. Superficially, it has a tibionavicular band running from the medal malleolus to the navicular. Adjacent to this are the fibers of the tibiocalcaneal ligament, and deep to these are the anterior and posterior tibiotalar ligaments, which run downward and obliquely away from each other from the lateral malleolus to the talus.

The anterior and posterior ligaments are thickenings of the joint capsule. The anterior ligament runs from the anterior tibia to the talus, while the posterior ligament is triangular in shape and runs from both the posterior tibia and fibula down to the posterior talus.

▶ **THE ANKLE**
Ligaments keep the joint stable and prevent excessive movement.

LIGAMENTS OF THE ANKLE

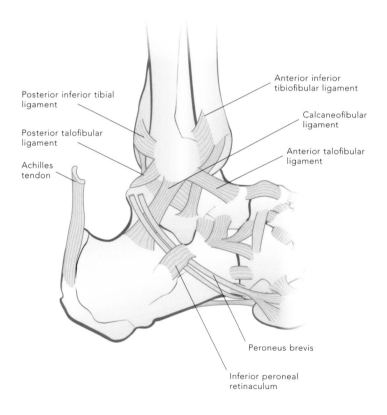

Posterior inferior tibial
ligament

Posterior talofibular
ligament

Achilles
tendon

Anterior inferior
tibiofibular ligament

Calcaneofibular
ligament

Anterior talofibular
ligament

Peroneus brevis

Inferior peroneal
retinaculum

Muscles of the ankle

Muscles of the front of the ankle joint

THE TIBIALIS ANTERIOR IS THE STRONGEST OF THE DORSIFLEXORS AND ORIGINATES FROM THE UPPER TWO-THIRDS OF THE LATERAL TIBIA. It narrows to a tendon at the front of the leg, crosses the ankle joint under the extensor retinaculum, and attaches on the medial cuneiform bone and the first metatarsal. When working in conjunction with tibialis posterior, it will invert the foot.

The peroneus longus originates from the superior and lateral surface of the fibula and the lateral tibial condyle. The muscle fibers converge into a tendon, which runs posteriorly to the lateral malleolus. This tendon then crosses under the foot and attaches to the medial cuneiform and base of the first metatarsal. The peroneus longus both everts and plantar flexes the foot.

The peroneus brevis muscle is deeper and shorter than the peroneus longus and originates on the inferior and lateral surface of the fibular shaft. The tendon descends with the peroneus longus tendon posterior to the lateral malleolus, passing over the calcaneus and the cuboid bones. The tendon then attaches to a tubercle on the fifth metatarsal. It everts and plantar flexes the foot. The peroneus tertius is a weak dorsiflexor and evertor of the foot; it arises from the lower anterior fibula and attaches on the fifth metatarsal.

▶ **ANKLE MUSCLES**

Many of the muscles of the lower leg converge at the ankle, thinning into tendons, which run across the front, back, and sides of the ankle joint.

MUSCLES OF THE ANKLE— ANTERIOR

Peroneus longus

Gastrocnemius

Extensor digitorum longus

Tibialis anterior

Superior extensor retinaculum

Soleus

Inferior extensor retinaculum

Peroneus longus tendon

Peroneus brevis

Superior extensor retinaculum

Achilles (calcaneal) tendon

Tibialis anterior

Extensor hallucis longus

Superior peroneal retinaculum

Extensor digitorum longus

Superior extensor retinaculum

Calcaneus

Tendon sheaths

Inferior extensor retinaculum

Tendon sheath

Extensor digitorum longus tendons

Extensor digitorum brevis tendons

Inferior peroneal retinaculum

Extensor hallucis longus tendon

Peroneus longus

Peroneus brevis

Extensor digitorum brevis

Peroneus tertius

Tuberosity of fifth metatarsal

Metatarsophalangeal joint

Muscles of the rear of the ankle joint

THE GASTROCNEMIUS AND THE SOLEUS ARE THE MAIN PLANTAR FLEXORS OF THE ANKLE. The gastrocnemius originates on the epicondyles of the tibia and inserts via the Achilles tendon onto the calcaneus. The soleus lies deeper and arises from the posterior surface of the upper one-third of the fibula and the posterior surface of the tibia. Its fibers converge to insert on the Achilles tendon. The gastrocnemius can exert maximal power at the ankle joint only when the knee is extended; however, the soleus is able to plantar flex from any knee position.

The plantaris is a long, thin muscle that originates from the lateral supracondylar ridge of the femur. It descends medially, converging into a tendon that runs between the gastrocnemius and soleus before blending with the Achilles tendon.

The tibialis posterior is the deepest out of the four muscles. It originates from the interosseous membrane between the tibia and fibula and the posterior surfaces of the two bones. The tendon runs posterior to the medial malleolus in a groove and attaches to the navicular and medial cuneiform bones. It inverts the foot and plantar flexes at the ankle.

▶ **MOVEMENT OF THE ANKLE**
The movements of the foot include dorsiflexion, plantar flexion, inversion or supination, and eversion or supination.

MUSCLES OF THE ANKLE—POSTERIOR

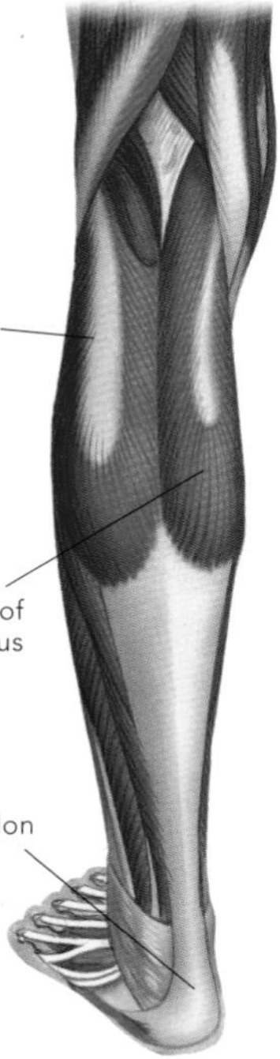

Lateral head of gastrocnemius

Medial head of gastrocnemius

Achilles tendon

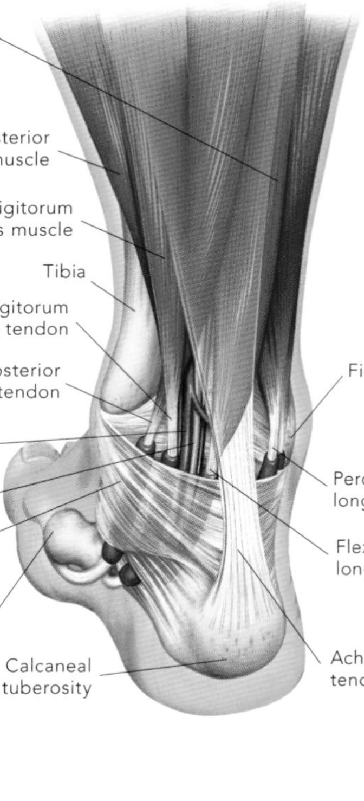

Flexor hallucis longus muscle

Tibialis posterior muscle

Flexor digitorum longus muscle

Tibia

Flexor digitorum longus tendon

Tibialis posterior tendon

Posterior tibial artery

Tibial nerve

Flexor retinaculum

First metatarsal

Calcaneal tuberosity

Fibula

Peroneus longus tendon

Flexor hallicus longus tendon

Achilles (calcaneal) tendon

Bones of the foot

THE BONES OF THE FOOT PROVIDE SUPPORT TO WITHSTAND THE WEIGHT OF THE BODY. They are divided into three groups: the tarsals, the metatarsals, and the phalanges.

The tarsal bones of the foot are organized into three rows: proximal, intermediate, and distal. The proximal tarsals are the talus and the calcaneus. They form the bony framework around the ankle and heel. The talus has three articulations: superiorly, it forms the ankle joint between the talus and the tibia and fibula; inferiorly, it forms the subtalar joint with the calcaneus; and anteriorly, it forms the talonavicular joint with the navicular. The calcaneus is a large, strong bone that transmits forces from the talus to the ground. It also articulates with the cuboid to form the calcaneocuboid joint.

The intermediate group contains just one bone: the navicular. This articulates with the talus posteriorly, the cuneiform bones anteriorly, and the cuboid bone laterally. The distal group contains four bones: the cuboid and the lateral, intermediate, and medial cuneiform bones.

Distal to the tarsals are the five metatarsal bones. Medially to laterally, they are numbered one to five, and each has a proximal base, a long shaft, and a distal head. Proximally, they articulate with the cuneiforms and cuboid bones to form the tarsometatarsal joint. At their base they articulate with their adjacent metatarsal to form the intermetatarsal joints. Distally, they each form the metatarsophalangeal joint with the proximal phalanx.

The phalanges are the bones of the toes. There are fourteen phalanges: a proximal, intermediate, and distal phalanx for each toe, with only a proximal and a distal phalanx for the big toe.

▶ **BONES OF THE FOOT**
The back of the foot, the tarsus, contains a group of seven bones, while the front of the foot contains the metatarsals and the phalanges.

CROSS SECTION THROUGH THE FOOT

THE FOOT

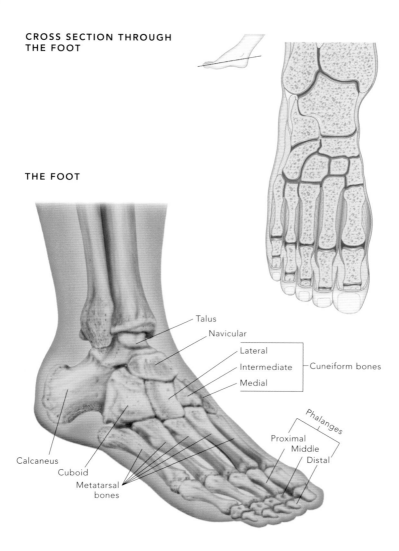

Talus

Navicular

Lateral
Intermediate — Cuneiform bones
Medial

Phalanges

Proximal
Middle
Distal

Calcaneus

Cuboid

Metatarsal bones

Ligaments and tendons of the foot

THE SUBTALAR JOINT IS SUPPORTED BY A NUMBER OF LIGAMENTS JOINING THE TALUS WITH THE CALCANEUS. The interosseous ligament plays an essential role in maintaining stability of the joint, and it is made up of two thick bands that occupy a central position. Other ligaments supporting the joint include the medial, posterior, and lateral talocalcanean ligaments, as well as the cervical ligament that provides strong support to the lateral side.

A network of ligaments supports the talocalcaneonavicular joint while limiting the inversion and eversion of the foot. One of the strongest and most significant of these is the plantar calcaneonavicular ligament. It runs from the talus to the entire width of the navicular. Due to its elasticity, it is often referred to as the "spring ligament" that helps maintain the arch of the foot during weight bearing.

An extensive network of ligaments and joint capsules supports the articulations between each of the tarsal bones. Between the tarsal bones and the second to fifth metatarsals are the tarsometatarsal joints, the joint spaces of which are also continuous with the intermetatarsal joints.

A joint capsule supports the metatarsophalangeal joints and is reinforced by ligaments and fibers from the extensor muscle tendons. Each joint has collateral ligaments that join on each side of the metatarsal, extending downward to the side of each phalanx. Upon becoming taut on flexion, they limit this action. Plantar ligaments attach to the plantar surface of the base of each phalanx. Similarly, collateral and plantar ligaments support proximal and distal interphalangeal joints.

▶ **LIGAMENTS OF THE FOOT**
Strong ligaments extend across much of the foot and ankle, providing stability and support.

LIGAMENTS OF THE FOOT

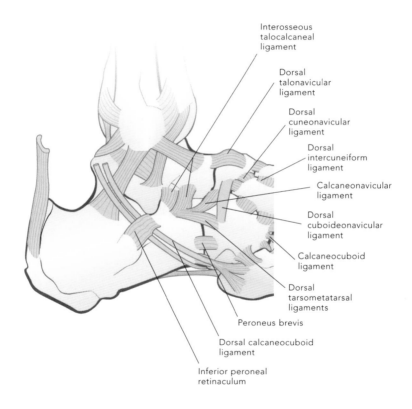

Interosseous talocalcaneal ligament

Dorsal talonavicular ligament

Dorsal cuneonavicular ligament

Dorsal intercuneiform ligament

Calcaneonavicular ligament

Dorsal cuboideonavicular ligament

Calcaneocuboid ligament

Dorsal tarsometatarsal ligaments

Peroneus brevis

Dorsal calcaneocuboid ligament

Inferior peroneal retinaculum

Muscles of the foot

MUSCLES THAT FLEX AND EXTEND THE TOES CAN BE DIVIDED INTO EXTRINSIC MUSCLES THAT ORIGINATE OUTSIDE THE FOOT AND INTRINSIC MUSCLES THAT ARE SOLELY LOCATED WITHIN THE FOOT.

The extensor digitorum longus lies deep to the tibialis anterior and originates from the lateral condyle of the tibia and the medial surface of the fibula. The muscle fibers converge into a tendon, which travels to the dorsal surface of the foot. The tendon splits into four and forms an extensor hood on each proximal phalanx, then attaches to the base of the distal phalanx of each toe. It produces extension of the lateral four toes and dorsiflexion of the foot.

The extensor hallucis longus is also an extrinsic muscle that originates from the medial surface of the fibular shaft. It attaches to the base of the distal phalanx of the big toe. Being the only extensor of the interphalangeal joint, it performs an important role during gait by helping the big toe clear the group as the leg swings through.

The extensor digitorum brevis lies deep to the tendon of the extensor digitorum longus. It originates from the calcaneus and the inferior extensor retinaculum and attaches to the extensor hood of the second, third, and fourth digits, as well as the proximal phalanx of the big toe. It assists in extending the medial four toes at the metatarsophalangeal and interphalangeal joints.

The lumbricals are four small muscles that produce flexion of the metatarsophalangeal joint and extension of the interphalangeal joints. They span between the flexor and extensor compartments of the foot between the metatarsals.

▶ **MOVEMENT OF THE FOOT**
There are no muscles in the phalanges of the toes, with the movement provided by tendons.

MUSCLES OF THE FOOT

Peroneus longus tendon

Peroneus brevis

Tibialis anterior

Extensor hallucis longus

Extensor digitorum longus

Achilles tendon

Superior extensor retinaculum

Superior peroneal retinaculum

Tendon sheaths

Calcaneus

Inferior extensor retinaculum

Tendon sheath

Extensor digitorum longus tendons

Extensor digitorum brevis tendons

Extensor hallucis longus tendon

Inferior peroneal retinaculum

Peroneus longus extensor

Extensor digitorum brevis

Peroneus brevis

Tuberosity of fifth metatarsal

Peroneus tertius

Metartarsal joint

Muscles of the foot *(cont.)*

The flexor digitorum longus is an extrinsic muscle that flexes the second to fifth toes. It originates from the medial surface of the tibia and attaches to the plantar surfaces of the base of the distal phalanges.

The flexor hallucis longus is a larger and stronger extrinsic muscle than the flexor digitorum longus. It produces the final push-off during gait. It arises from the posterior surface of the fibula and attaches to the plantar surface of the distal phalanx of the big toe.

The flexor accessorius originates on the calcaneus and inserts onto the tendon of flexor digitorum longus. Contraction of the flexor accessorius assists the flexor digitorum longus muscle to flex the toes, thus straightening its line of pull toward the calcaneus rather than toward the medial malleolus.

The flexor digitorum brevis lies in the center of the foot between the plantar aponeurosis and the tendons of the flexor digitorum longus. It arises from the calcaneus and the plantar aponeurosis and attaches to the base of the middle phalanges of the lateral four digits. It flexes the second through fourth digits at the proximal interphalangeal joints.

Both the little toe and the big toe have their own flexor muscles: the flexor digiti minimi brevis and the flexor hallucis brevis, respectively. The flexor hallucis brevis muscle originates from the plantar surfaces of the cuboid and lateral cuneiform bones and from the tendon of the posterior tibialis tendon. It inserts on the base of the proximal phalanx of the big toe. The flexor digiti minimi brevis muscle originates from the base of the fifth metatarsal and inserts at the base of the proximal phalanx of the fifth digit.

The plantar and dorsal interossei arise from the sides of the metatarsals and attach to the proximal phalanges. Both of these muscle groups flex the metatarsophalangeal joints with the plantar interossei, also adducting digits three to five and the dorsal interossei abducting digits two to four.

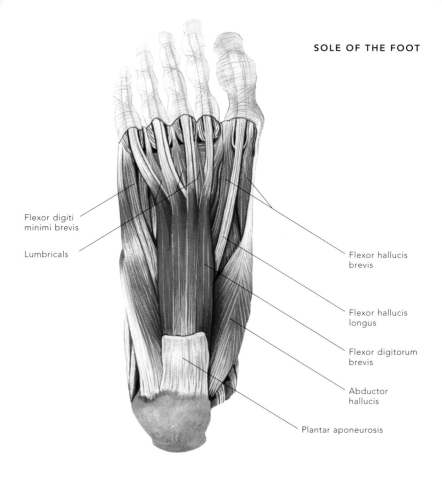

Flexor digiti
minimi brevis

Lumbricals

Flexor hallucis
brevis

Flexor hallucis
longus

Flexor digitorum
brevis

Abductor
hallucis

Plantar aponeurosis

▲ **MOVEMENT OF THE
FOOT**

The tendons of the flexor
muscles provide flexion of
the toes.

Cycling

Each sport has its own specific movements, requiring the activation of specific muscles that need to be coordinated to contract at a specific time. Even movements that require the same muscles can have very different patterns of coordination, so that training for one sport cannot fully transfer benefits to another. Running, for example, is quite different from cycling.

In cycling, the motion can be described as circular, where 0° is the top of the pedal stroke and 180° is with the pedal at the bottom. The drive phase begins immediately after the pedal has reached 0° for one of the legs. At this point, the hip for this leg is flexed at almost 90° and the ankle is dorsiflexed. The hip and knee extensors are the primary muscle groups involved in driving the pedal down. The vastus medialis and vastus lateralis contract first just before 0°, followed by the gluteus maximus

shortly after. At around 45°, the hamstring muscles of the semimembranosus and the long head of the biceps femoris contribute to hip extension. Both heads of the gastrocnemius and soleus then begin to contract to plantar flex the ankle joint. Some external rotation of the hip is also produced in order to contribute to a maximum force generated by contraction of the gluteus medius, which is assisted by anterior fibers of the gluteus maximus.

▶ **DRIVE PHASE**
The hip and knee extensors of the right leg initiate the drive phase.

Cycling (cont.)

At around 135°, the knee extensors and gluteus maximus stop contracting powerfully, with the hamstring muscles continuing hip extension to 180°. In order to maximize drive through the pedals, the flexor digitorum longus, flexor hallucis longus, and intrinsic flexors contract to drive through the toes.

The recovery phase from 180° back to 0° is mostly passive and generated by the drive phase of the other leg. At the bottom of the pedal stroke, the knee and hip reach their maximum extension during the cycling motion, with the knee remaining at about 30 degrees of flexion. The gastrocnemius and the hamstring muscles continue to contract beyond 180° to contribute to knee flexion, pulling the pedal through the bottom of the stroke and up into the initial recovery phase. The tibialis anterior will then contract at around 225° to bring the ankle back into dorsiflexion until 0°. The rectus femoris begins to contract at 270° along with the other hip flexors, iliopsoas, sartorius, and tensor fasciae latae. The rectus femoris will continue to contract past 0° through the drive phase, joined by the other quadriceps muscles.

▶ **RECOVERY PHASE**

With the left leg in its drive phase, the right leg is in the recovery phase. The gastrocnemius and hamstring muscles enable knee flexion.

Chapter 11:
The spine

The spine consists of a series of small bones or vertebrae, separated by cartilaginous discs. The spine has an essential role to play in determining the body's posture, and crucially houses and protects the spinal cord that transmits nerve impulses to and from the brain. Damage to the spine can be both painful, such as when a disc moves, and life-changing, if the spinal cord is damaged.

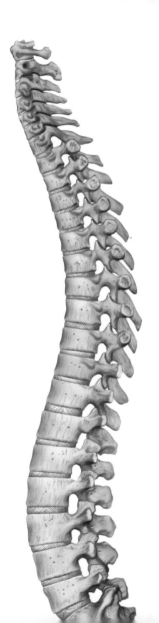

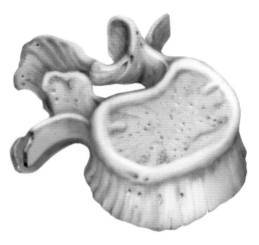

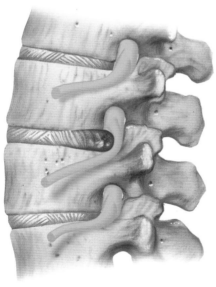

Overview of the spine

Properties of the spine

The spine, otherwise known as the vertebral column, is composed of vertebrae and intervertebral discs. It sits in the midline of the back, from the cranium to the apex of the coccyx. The adult spine consists of 24 articulating vertebrae and nine fused vertebrae that are anatomically separated into five regions: seven cervical (C), twelve thoracic (T), five lumbar (L), five sacral (S), and four coccygeal (Co). The spine is a vital part of the skeletal system that protects the spinal column, supports the weight of the body above the pelvis, and provides both a flexible axis for the body to move and a rigid axis for the attachment of muscles and ligaments.

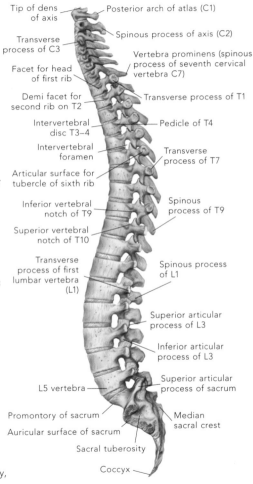

Tip of dens of axis — Posterior arch of atlas (C1)

Transverse process of C3

Spinous process of axis (C2)

Facet for head of first rib

Vertebra prominens (spinous process of seventh cervical vertebra C7)

Demi facet for second rib on T2

Transverse process of T1

Intervertebral disc T3–4

Pedicle of T4

Intervertebral foramen

Transverse process of T7

Articular surface for tubercle of sixth rib

Inferior vertebral notch of T9

Spinous process of T9

Superior vertebral notch of T10

Transverse process of first lumbar vertebra (L1)

Spinous process of L1

Superior articular process of L3

Inferior articular process of L3

L5 vertebra

Superior articular process of sacrum

Promontory of sacrum

Auricular surface of sacrum

Median sacral crest

Sacral tuberosity

Coccyx

▶ **THE SPINE**
The spine provides stability, flexibility, and a wide range of movement.

The adult spine contains four curves, which give it an S-shaped appearance. The thoracic and sacral regions are convex curves (kyphotic curvature), whereas the cervical and lumbar regions are concave curves (lordotic curvature). The thoracic and sacral curvatures are primary curves that develop during the fetal period and are retained throughout life. The cervical and lumbar lordotic curvatures develop during infancy and are known as secondary curves. The function of kyphotic and lordotic curves is to provide balance, flexibility, and absorption of compressive forces.

Movement of the spine

One of the main functions of the spine is to provide flexibility and movement. Spinal anatomy is defined by the structure of the vertebrae, the facet (or zygapophysial) joints, and the intervertebral discs that are present between the vertebrae. Facet joints are present in the posterior region of each vertebra and allow the spine to rotate and move in different directions. The intervertebral discs are fibrocartilaginous discs that absorb the stress incurred during movement. In the thoracic region of the spine, the rib cage attachments provide increased stability at the expense of spinal range of movement. The cervical region of the spine has the greatest amount of flexibility, with flexion, extension, and rotation all possible. Within the lumbar region of the spine is a similar range of movement, with significant contribution to spinal flexion and extension but not to rotational motion. There is limited motion in the lower nine vertebrae, as the five sacral and four coccygeal regions are fused to form the sacrum and coccyx, respectively. The only articulation is between the last lumbar vertebra (L5) and the first sacral vertebra (S1).

Joints

The spine consists of several types of joints that provide flexibility, support, and movement of the spine. The joints between the vertebral bodies are known as symphyses (cartilaginous joints). The vertebral bodies articulate with each other to provide strength and allow for weight bearing, and are bound together by IV discs, providing a semi-rigid column that is supported by ligaments. The joints of the vertebral arches are known as zygapophysial (facet) joints. Articulations occur between the inferior and superior articular processes of corresponding vertebrae, thus allowing a gliding movement. These synovial plane joints also have a thin joint capsule that provides nutrients and oxygen for the joint.

Further classifications of spinal joints include the costovertebral joint, the craniovertebral joints—which is termed the atlantooccipital and atlantoaxial joints—and the sacroiliac joint. The costovertebral joint is the articulation between the thoracic vertebrae and the ribs, which are strengthened by several ligaments that add increased stability but limit motion. The craniovertebral joints permit movement between the cervical vertebrae and the cranium, whereas the sacroiliac joints are the joints between the left and right ilia of the pelvis and the sacrum.

Ligaments

The intervertebral articulations of the spine are supported by strong accessory ligaments, including the ligamentum flavum, which joins laminae of adjacent vertebrae; the interspinous ligament, which connects the root and apex of the spinous processes; the supraspinous ligament, which joins adjacent spinous processes; the ligamentum nuchae, which joins the external occipital protuberance and posterior border of the foramen magnum to the spinous processes of the cervical vertebrae; and the intertransverse ligament, which connects the adjacent transverse processes. The ligamentum nuchae runs from C1 to C7, providing a surface area for muscular attachment, and is separate from the supraspinous and interspinous ligaments.

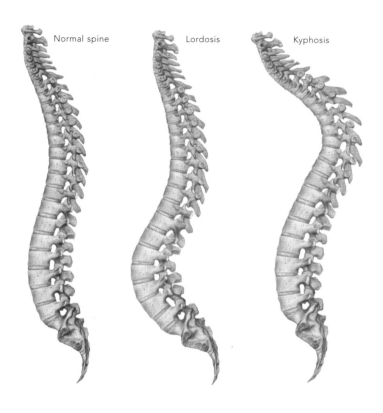

Normal spine   Lordosis   Kyphosis

▲ SPINE CURVATURE

Lordosis curvature sees the pelvis aligned with the top of the thighs. Kyphosis curvature affects the cervical, thoracic, and sacral regions of the spine.

Vertebrae

All vertebrae are structurally similar, with variations in size and regional characteristics. The basic structure of a typical vertebra is as follows:

- Vertebral body – This is the largest feature of the vertebra; it is cylindrical in shape and sits anteriorly. It gets progressively larger as it descends the spinal column due to its weight-bearing capacity.

- Vertebral arch – This sits posteriorly to the body and consists of a pedicle and laminae on both sides (right and left). A pedicle is a short cylindrical projection that leaves the vertebral body posteriorly and joins with a flat plate of bone, known as the lamina. Pedicles vary in size and direction of projection, depending on which region they belong to.

- Vertebral foramen – The vertebral arch and body combine to form the vertebral foramen. Collectively, the foramen of each articulating vertebra forms the vertebral canal, which houses the spinal cord.

- A spinous process – A single posterior projection, typically inferiorly, from the laminae of the vertebral arch. The spinous processes are usually palpable, often revealed down the midline of the back during flexion.

- Two transverse processes – A pair of posterolateral projections from the vertebral arch (junction of the pedicle and laminae). Along with the spinous process, the transverse processes provide an attachment site for the deep back muscles, acting as levers throughout each spinal segment.

▶ **VERTEBRAE**

Each vertebra type shares some structural characteristics with the others, although each has specific features relevant to its location and purpose in the spine. Shown opposite, a lumbar vertebra.

- Four articular processes –
  Two superior and two inferior
  processes that arise from the
  vertebral arch. All four contain
  articular surfaces (facets) that act
  as the site of articulation with
  vertebrae above and below
  (forming zygapophysial joints).

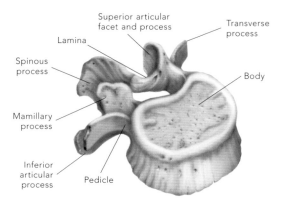

Regions of the spine

Introduction to the spine

Each region of the vertebral column has structural characteristics that separate it from other regions. In addition, some individual vertebrae have structural variations that identify their location within the spine. For example, the C7 vertebra has the longest spinous process, making it easily palpable—acting as a regional landmark—when examining the spine.

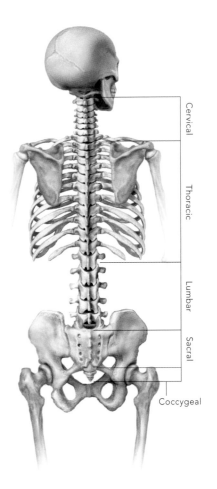

Cervical

Thoracic

Lumbar

Sacral

Coccygeal

▶ **WEIGHT-BEARING CAPABILITIES**

The design of the spine creates load-bearing capabilities to carry the weight of the head and upper body.

Cervical spine

The cervical vertebrae are the smallest in the spinal column. For example, vertebrae C3–C7 have a small but broad vertebral body and a large, triangular foramen. The spinous processes of the cervical spine are short and bifid, with the most distinguishable feature being the oval-shaped transverse foramen that allows the passage of vertebral arteries. Small in size, the cervical vertebrae have the greatest range of motion due to their thick intervertebral discs, and the orientation of the articular facets (superior and posterior). The first two vertebrae, C1 and C2, are atypical and are referred to as the atlas and axis, respectively. The atlas is without a body or a spinous process, and this ring-shaped vertebra is responsible for supporting the cranium. The axis is the strongest vertebra in the cervical spine and it is the point where C1 rotates the head.

Lumbar spine

The largest vertebral bodies are present in the lumbar spine and these provide an important weight-bearing role. The lumbar vertebral bodies are shaped like a kidney: concave posteriorly and larger side to side than they are front to back. They have a triangular foramen that is larger than in the thoracic vertebrae but smaller than in the cervical vertebrae. Lumbar vertebrae have short and thick spinous processes and long and thin transverse processes. The articular processes extend vertically, and the facets are sagitally oriented, allowing all movements except for rotation. The lumbar vertebra L5 is the largest vertebra in the spine and is wedge-shaped due to the larger anterior portion.

Discs

INTERVERTEBRAL DISCS SERVE AS STRONG ATTACHMENT POINTS FOR CORRESPONDING VERTEBRAL BODIES, thus permitting movement at each segment and helping absorb shock. Each intervertebral disc contains a fibrous outer ring—the annulus fibrosus—which surrounds an inner gelatinous mass, known as the nucleus pulposus.

The annulus fibrosus is securely attached to the vertebral body via the epiphysial rims. It is made up of tough layers of fibrocartilage running obliquely from one vertebra to another. The nucleus pulposus consists primarily of water (80–90%), allowing it to deform when stretched or compressed. This gives the disc the ability to absorb shock and permit movement of the spine. Throughout the spinal column (C3–S2), discs vary in size and thickness—the largest discs are present in the cervical and thoracic regions.

▸ **SHOCK ABSORPTION**
Intervertebral discs are durable shock absorbers within the spine.

INTERVERTEBRAL DISC

Annulus
fibrosus

Nucleus
pulposus

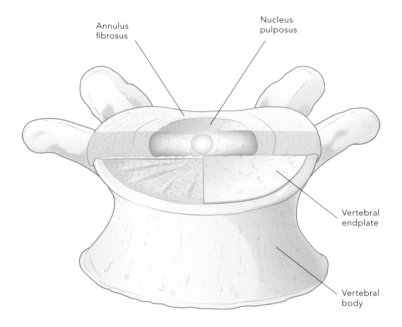

Vertebral
endplate

Vertebral
body

Surface and deep muscles of the back

THE MUSCLES OF THE BACK CAN BE CATEGORIZED AS SUPERFICIAL (EXTRINSIC) OR DEEP (INTRINSIC). Superficial muscles connect the spine with the pectoral girdle and humerus, thus controlling movement of the upper limbs. These muscles include the trapezius, latissimus dorsi, levator scapulae, and rhomboids. All of these muscles receive innervation from the anterior rami of the cervical spine, with the trapezius receiving its innervation from the spinal accessory nerve (cranial nerve XI). Lying deep to these muscles is the serratus posterior, which is smaller and thinner in nature. The serratus posterior is innervated by the intercostal nerves and assists with respiration, while lying deep to both the rhomboids and latissumus dorsi.

The deep muscles of the back are responsible for controlling movement of the spine and maintaining posture. The splenius muscles are the most superficial of the intrinsic muscles, and these originate from the spinous processes of C7–T6 and attach to the cervical vertebrae and cranium, the splenius cervicis and capitis, respectively.

Lying deeper to the splenius muscles are the three erector spinae muscles. Positioned from medial to lateral, they are the iliocostalis, longissimus, and spinalis erector spinae muscles. These muscles are located between the spinous and transverse processes of the spine and are covered by fascia in the lumbar and thoracic regions (thoracolumbar fascia). Deep to the erector spinae are the transversospinalis muscles (multifidus, semispinalis, and rotatores). These muscles are small, with a high density of muscle spindles that allow for fine movement. They originate from the transverse processes and attach to the spinous process of the vertebra above.

▶ **SUPPORT AND MOVEMENT**
The strong muscles of the back stabilize the area, providing support and movement to the spinal column.

SUPERFICIAL

DEEP

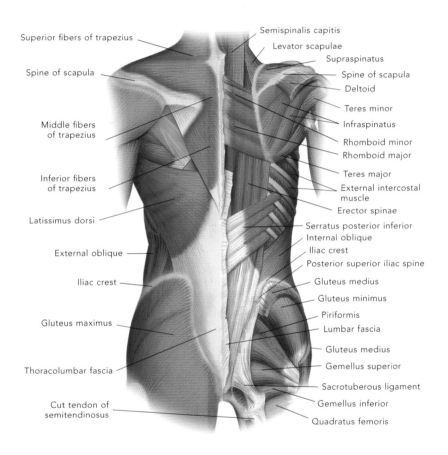

Superior fibers of trapezius

Spine of scapula

Middle fibers of trapezius

Inferior fibers of trapezius

Latissimus dorsi

External oblique

Iliac crest

Gluteus maximus

Thoracolumbar fascia

Cut tendon of semitendinosus

Semispinalis capitis

Levator scapulae

Supraspinatus

Spine of scapula

Deltoid

Teres minor

Infraspinatus

Rhomboid minor

Rhomboid major

Teres major

External intercostal muscle

Erector spinae

Serratus posterior inferior

Internal oblique

Iliac crest

Posterior superior iliac spine

Gluteus medius

Gluteus minimus

Piriformis

Lumbar fascia

Gluteus medius

Gemellus superior

Sacrotuberous ligament

Gemellus inferior

Quadratus femoris

Injuries and slipped discs

DISC HERNIATION (SLIPPED DISC) IS A COMMON INJURY THAT OCCURS AT THE SPINE, OFTEN AS A RESULT OF DISC DEGENERATION. The ability of the nucleus palposus to retain water decreases with age, making the discs less flexible and exposing them to a higher risk of injury. A herniation occurs when the nucleus palposus protrudes into, or out of, the annulus fibrosus. Herniations are commonly posteriorlateral, often encroaching on the spinal nerves and causing referred pain. Sciatica is a herniation of the L5–S1 disc, compressing the nerve root to produce a radicular pain down the buttocks into the back of the leg.

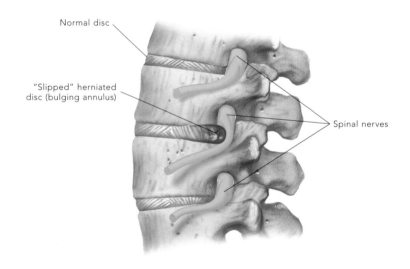

Normal disc

"Slipped" herniated disc (bulging annulus)

Spinal nerves

SLIPPED DISC

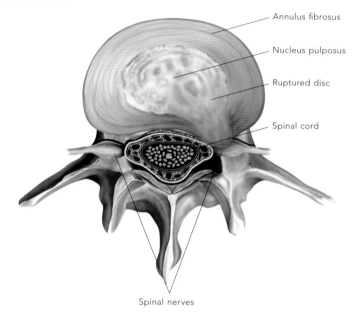

Annulus fibrosus

Nucleus pulposus

Ruptured disc

Spinal cord

Spinal nerves

▲ **SLIPPED DISC**

If the nucleus pulposus intrudes into or out of the annulus fibrosus, it can encroach and irritate a spinal nerve.

Chapter 12:
Head and cervical spine

The skull and neck support the body's methods of communication, respiration, and nutrition. The skull also houses and protects the human brain with sturdy rigid bones, while facilitating the strong, powerful movements that are needed to eat food. The structure of the neck enables a range of movements in different planes while providing support and protection to the thorax and throat.

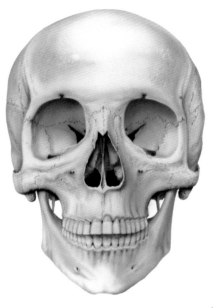

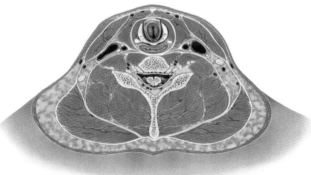

Skull

The skull consists of a number of bones, including eight cranial bones and fourteen facial bones. The cranial bones surround the brain, providing protection. Generally, the facial bones are attachment points for muscles and provide structures such as the eye socket for the eye. The orbit, which is the eye socket, is composed of seven bones, including the medial orbital wall—or lacrimal bone—one of the most fragile bones in the face.

The skull has particular points that are thickened and therefore resistant to forces placed on them. They are thickened into pillars—or buttresses—and transmit the force away from the more vulnerable parts of the skull, thus helping to resist fracture. Buttresses can be found around the orbit, the mouth, the chin, and at the occiput.

SKULL—SIDE

External acoustic meatus
Parietal bone
Frontal bone
Temporal line
Supraorbital notch
Ethmoid bone
Lacrimal bone
Nasal bone
Zygomatic bone
Zygomatic process of temporal bone
Pterygoid process of sphenoid bone
Maxilla
External occipital protuberance
Occipital bone
Coronoid process of mandible
Mastoid process
Styloid process
Tympanic plate of temporal bone
Condylar process of mandible
Mandibular notch
Body of mandible
Wisdom tooth
Ramus of mandible
Angle of mandible

SKULL—FRONT

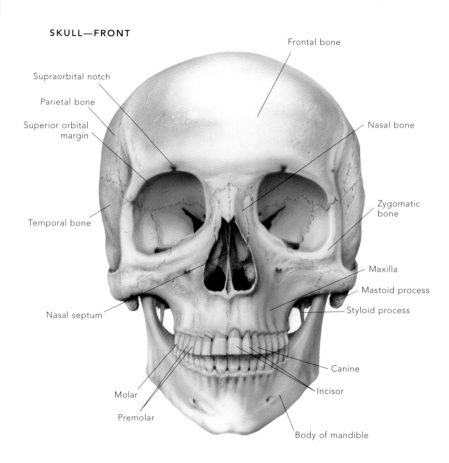

Supraorbital notch

Parietal bone

Superior orbital margin

Temporal bone

Nasal septum

Molar

Premolar

Frontal bone

Nasal bone

Zygomatic bone

Maxilla

Mastoid process

Styloid process

Canine

Incisor

Body of mandible

▲ BONES OF THE FACE

The face consists of the frontal bone, zygoma, maxilla, and the mandible.

Skull *(cont.)*

Immovable joints, held together by fibrous connective tissue, connect the bones of the adult skull. The exception is the temporomandibular joint (or the jaw), which is a synovial joint. This is to enable motion at the joint and is how humans are able to open and close their mouth. The joints of the skull are relatively mobile at birth and grow more rigid through aging. This enables greater protection of the brain and is why a baby's head can sometimes appear to be formed differently from an adult's head.

The bones that make up the nose are the vomer and ethmoid, which have bony cavities known as sinuses. Sinuses may be filled with various structures, such as glands or blood. The flow of blood through the sinuses may be disrupted due to infection that may be experienced when someone is suffering from a cold or flu, and it is evident by trouble with breathing as well as a feeling of congestion around the nose.

Within the skull are numerous foramina (passageways) that provide a protected passage for structures, including blood vessels, nerves, and muscles. Within a relatively small space are a vast number of vessels and nerves that must provide adequate supply to the structures of the head. Failure to do so could result in serious consequences that include a loss of the senses.

▶ **COMPLEXITY**

The combination of its bones makes the skull one of the most complex structures in the human body.

SKULL—CROSS SECTION

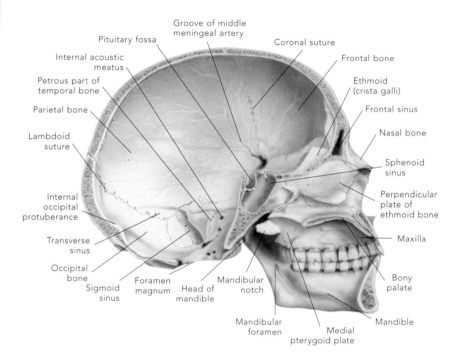

Groove of middle meningeal artery

Pituitary fossa

Internal acoustic meatus

Coronal suture

Petrous part of temporal bone

Frontal bone

Parietal bone

Ethmoid (crista galli)

Lambdoid suture

Frontal sinus

Nasal bone

Internal occipital protuberance

Sphenoid sinus

Transverse sinus

Perpendicular plate of ethmoid bone

Occipital bone

Maxilla

Sigmoid sinus

Foramen magnum

Head of mandible

Mandibular notch

Bony palate

Mandibular foramen

Medial pterygoid plate

Mandible

Skull *(cont.)*

The base of the skull is formed by the occipital bone, and within this is the foramen magnum. This is the passageway for the spinal cord to enter into the brain, and therefore must provide sufficient space for the spinal cord and lower brainstem to transmit through safely. Different parts of the skull are shaped to support specific parts of the brain; for example, the anterior cranial fossa supports the frontal lobes of the cerebrum. The external surface of the base of the skull provides attachment points for the posterior cervical musculature. In addition, there are large occipital condyles that articulate with the facets of the atlas or first cervical vertebra.

The skull is a complex combination of bones fused together to form a protective shield for the brain. There are numerous structures contained within the head that sit within a relatively small space. Blunt trauma to the head or face may result in damage to any of these structures, which must be considered if assessing anyone following this type of injury.

▶ **THE BASE OF THE SKULL**
The image shows the complex arrangement at the base of the skull for important structures, and the main bony prominences.

BASE OF THE SKULL

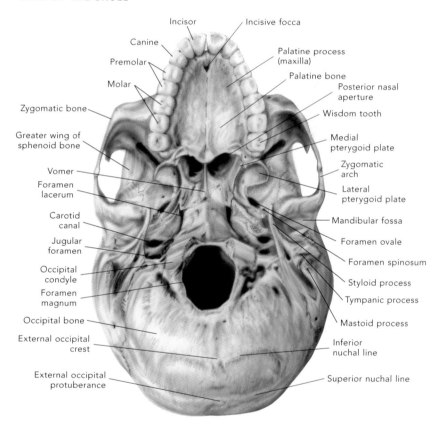

Incisor

Incisive focca

Canine

Premolar

Palatine process (maxilla)

Molar

Palatine bone

Posterior nasal aperture

Zygomatic bone

Wisdom tooth

Greater wing of sphenoid bone

Medial pterygoid plate

Vomer

Zygomatic arch

Foramen lacerum

Lateral pterygoid plate

Carotid canal

Mandibular fossa

Jugular foramen

Foramen ovale

Occipital condyle

Foramen spinosum

Foramen magnum

Styloid process

Occipital bone

Tympanic process

External occipital crest

Mastoid process

Inferior nuchal line

External occipital protuberance

Superior nuchal line

The neck

Introduction to the neck

The neck is referred to as the cervical spine and is made up of seven cervical vertebrae. This number is consistent across the population, although on occasion someone may present with C1 and C2 fused to the occiput. The model cervical spine will have a curve, and this is termed cervical lordosis, meaning that the convexity faces anteriorly. These curves form at approximately 3 months of age after the postural reflexes have been developed.

As with the other regions of the spine, there is an intervertebral disc, facet joint, and ligaments between adjacent vertebrae. This is termed a motion segment—when motion occurs in the cervical spine, it is due to a combination of movements of motion segments. Each vertebra is secured in place by ligaments, and there is a fibrous capsule around the facet joint. Each vertebra has a vertebral foramen through which the spinal cord passes. On each side of the vertebra are intervertebral foramina, which transmit the spinal nerves along with their coverings and vessels.

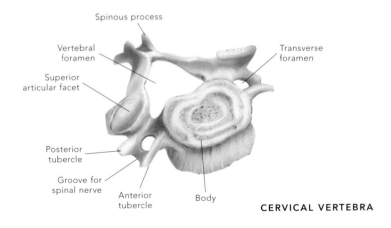

Spinous process

Vertebral foramen

Transverse foramen

Superior articular facet

Posterior tubercle

Groove for spinal nerve

Anterior tubercle

Body

CERVICAL VERTEBRA

The cervical vertebrae are smaller than other vertebrae and the cervical facets are also oriented horizontally in comparison to other spinal regions. The shape of these bones allows for significant degrees of flexion and extension, as well as rotation throughout the entire cervical spine, although movement at individual motion segments is relatively small. The spinous process of C7 is an easily palpable structure that can be used as an orientation point for other structures to be palpated.

▼ **THE CERVICAL VERTEBRAE**
Owing to their shape, the cervical vertebrae provide a large degree of flexibility.

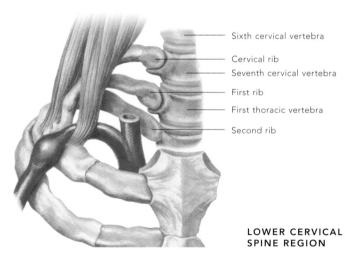

Sixth cervical vertebra

Cervical rib
Seventh cervical vertebra

First rib

First thoracic vertebra

Second rib

LOWER CERVICAL SPINE REGION

The atlas and axis

THE JOINT BETWEEN THE SKULL AND THE SPINAL COLUMN IS KNOWN AS THE ATLANTOOCCIPITAL JOINT. It is made up of the occipital condyles of the skull that sit on the articular fossa of the atlas (C1), thus allowing flexion and extension to occur. The atlas sits on the axis (C2) to form the atlantoaxial joint. Most of the rotation in the cervical spine occurs at the atlantoaxial joint, which is known as a trochoid, or pivot, joint. In contrast, the other joints in the cervical spine are arthrodial, or gliding, joints.

These upper cervical vertebrae have transverse processes that have foramina (openings) that permit the vertebral arteries to pass through and reach the brainstem. These vessels can become damaged with extreme rotation of a hyperextended neck, which may affect the blood flow to the brain.

▶ **ATLANTOAXIAL JOINT**
The atlas joint differs slightly from the other joints in the cervical spine, as can be seen in the image.

ATLAS AND AXIS

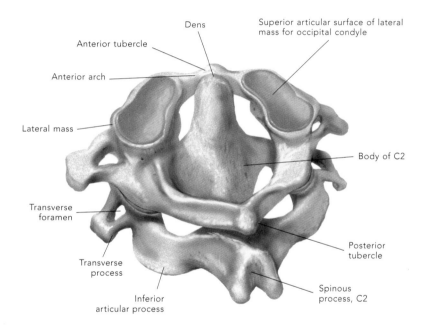

Dens

Superior articular surface of lateral
mass for occipital condyle

Anterior tubercle

Anterior arch

Lateral mass

Body of C2

Transverse
foramen

Posterior
tubercle

Transverse
process

Inferior
articular process

Spinous
process, C2

Muscles of the face

Introduction to the muscles of the face

THE FACE IS MADE UP OF MANY MUSCLES THAT ARE RESPONSIBLE FOR THE EXPRESSIONS THAT HUMANS CAN MAKE. In general, these facial muscles are thin, flat bands that attach to facial bones or cartilage and then attach to the skin. They also attach to the fibrous tissue that envelops the sphincter muscles of the orbit or mouth. The muscles responsible for facial expressions are innervated mainly by the seventh cranial nerve (VII), which is also involved in the sensation of taste.

The muscles are divided into groups based on the region in which they are located. These groups are the epicranial group, which moves the scalp; the orbital group, which moves the muscles around the eye; the nasal group, which moves the nose; the oral group, which moves the mouth; and the auricular muscles, which move the ears.

Facial muscles can be referred to as those that make us cheery and those that make us sad. For example, orbicularis is involved in movement around the eye, and zygomaticus major is involved in motion around the cheekbone when we are happy. Examples of "sad" muscles include the frontalis and depressor anguli oris.

The orbicularis oculi and orbicularis oris are sphincter (ring-shaped) muscles that have specific roles, namely, closing the eyelids and tightening the lips. The nasalis can compress and dilate, influencing the size of the nasal openings; this is demonstrated when flaring the nostrils.

▶ **THE FACE**

A multitude of muscles covers the facial area, from the very small muscles in the forehead to the circular muscles of the mouth and eye areas.

MUSCLES OF THE FACE—SIDE

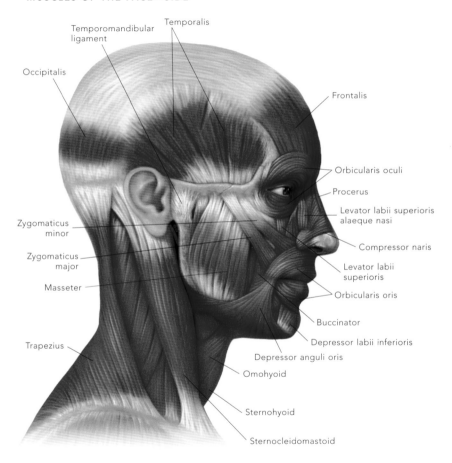

Temporalis

Temporomandibular ligament

Occipitalis

Frontalis

Orbicularis oculi

Procerus

Levator labii superioris alaeque nasi

Compressor naris

Levator labii superioris

Orbicularis oris

Zygomaticus minor

Zygomaticus major

Masseter

Buccinator

Depressor labii inferioris

Trapezius

Depressor anguli oris

Omohyoid

Sternohyoid

Sternocleidomastoid

Deep and surface muscles of the face

THE MUSCLES IN THE FACE ARE ARRANGED IN LAYERS, WITH SOME LOCATED DEEP AND SOME SUPERFICIAL. These superficial muscles are attached to the skin and are responsible for creating facial movements. The buccinator muscle is one of the deeper muscles that assists in the exhalation of air from the mouth. The procerus muscle extends upward from the bridge of the nose and is responsible for eyebrow movement. Although the muscles may be defined as deep or superficial, there is little difference between them in their layering and other tissues such as fascia and cartilage. Muscles in the scalp, such as the temporalis, are termed as deep muscles and form a protective layer over the skull. They are also responsible for movement and connect to other muscles via fascial attachments.

▶ **MUSCLES OF THE MOUTH**

The circular muscles of the lips form the mouth's muscular opening and participate in speech.

DEEP AND SURFACE MUSCLES
OF THE FACE

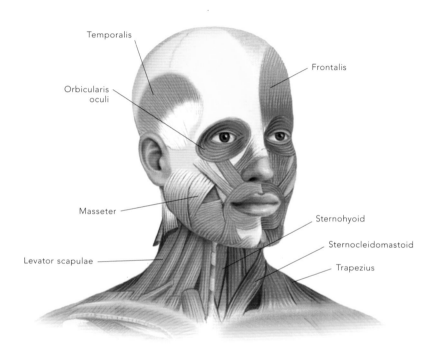

Temporalis

Frontalis

Orbicularis
oculi

Masseter

Sternohyoid

Sternocleidomastoid

Levator scapulae

Trapezius

Muscles of the jaw

THE TEMPOROMANDIBULAR JOINT OF THE JAW IS RESPONSIBLE FOR MASTICATION, AND THE MUSCLES INVOLVED PRIMARILY MOVE THE MANDIBLE THROUGH ELEVATION, depression, protrusion, retraction, and lateral motion. Although there are two joints, they need to work together to create smooth movement of the mandible. Chewing in particular is produced by the temporalis and masseter muscles that elevate on one side while the lateral pterygoid muscle on the opposite side contracts.

These jaw muscles may be contracted without the individual realizing it, such as when someone is stressed and clenches their teeth. They may also be involved when an individual grinds their teeth in their sleep. Overactivity of these muscles can lead to headaches in the bitemporal and preauricular regions since they are in a relatively constant state of tone, meaning that

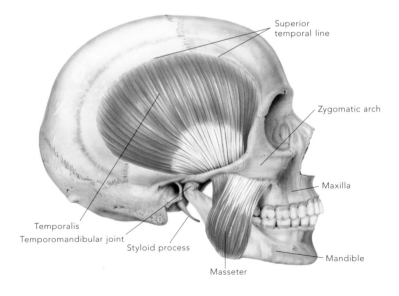

Superior temporal line

Zygomatic arch

Maxilla

Temporalis

Temporomandibular joint

Styloid process

Masseter

Mandible

they remain partially contracted. This can create tension at the muscle attachment points. The medial and lateral pterygoid muscles may become restricted, thus leading to decreased motion of the temporomandibular joint. This can be linked to pain or restriction in the cervical spine due to the proximity of the structures and the potential influence of the muscles in the area.

▼ **MOVEMENT OF THE JAW**

The muscles around the jaw permit a wide range of movements to the mandible.

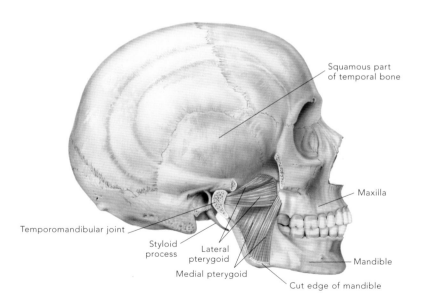

Squamous part of temporal bone

Maxilla

Temporomandibular joint

Styloid process

Lateral pterygoid

Medial pterygoid

Mandible

Cut edge of mandible

Muscles of the neck

Introduction to muscles of the neck

THE NECK IS A COMPLEX REGION WITH MULTIPLE STRUCTURES TRANSMITTING THROUGH IT. Within the neck are muscles that lie both deep and superficially and often have multiple roles. Although the muscles are responsible for movement, they also play an important role in supporting the weight of the head and providing strength and stability to the cervical spine and the visceral column. In general, the muscles can be divided according to their location: anterior, posterior, or lateral. Starting with the lateral muscles that can be seen in the image on the opposite page, the sternocleidomastoid provides a dividing line between the anterior and lateral groups, forming them into triangular areas. The sternocleidomastoid itself acts to flex the neck laterally when working unilaterally and to flex the neck when working bilaterally. The image displays the neck muscles from a lateral view, and the scalenes are just posterior to the sternocleidomastoid. The scalene muscles originate in the posterior region of the cervical spine and orientate toward the ribs on the anterior aspect. Their muscle action is lateral flexion to the same side, but they are also involved in breathing and assist in elevation of the rib cage during inspiration.

▶ **MOVEMENT OF THE UPPER SPINE AND HEAD**

The range of movement available in the cervical spine is provided by musculature at the front and back of the neck, along with muscles on both sides.

MUSCLES OF THE NECK

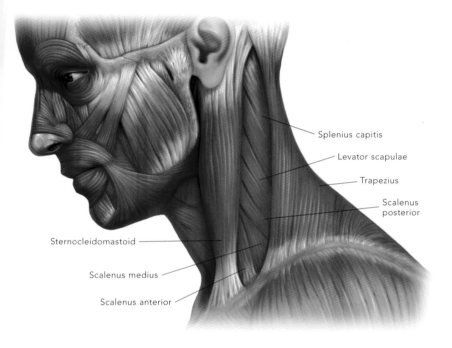

Splenius capitis

Levator scapulae

Trapezius

Scalenus posterior

Sternocleidomastoid

Scalenus medius

Scalenus anterior

Deep and surface muscles of the neck

THE ANTERIOR REGION OF THE NECK ALSO HAS SUPERFICIAL AND DEEP MUSCLES. The main muscles in the anterior region are known as the platysma, sternocleidomastoid, suprahyoid, and infrahyoid. These muscles have a responsibility for movement of the hyoid bone and play a role in actions such as swallowing and making sounds linked to larynx movement.

The posterior region of the neck has several muscles that are covered by a layer of cervical fascia. The most superficial muscle is the upper fibers of the trapezius muscle, which has numerous functions related to neck motion, including lateral flexion and rotation. Deep to the trapezius are the splenius muscles that extend and rotate the neck. Other muscles of the neck include the erector spinae longissimus, the spinalis, the transversospinalis, and the suboccipital muscles. The smaller muscles are termed the intrinsic movers and function as stabilizers for motion segments where required. If the deeper muscles are not functioning appropriately, the more superficial muscles will compensate, which often leads to overuse. Therefore, these muscles may become hypertonic—meaning there is too much muscle tone—due to excessive contraction in the muscle, even during resting. This can lead to other problems, such as tension headaches and pain in the cervical spine.

▶ **HEAD AND SPINE SUPPORT**
Strong muscles surround the neck region, providing support for the head and strengthening the cervical vertebrae.

DEEP AND SURFACE MUSCLES
OF THE NECK

SURFACE MUSCLES

DEEP MUSCLES

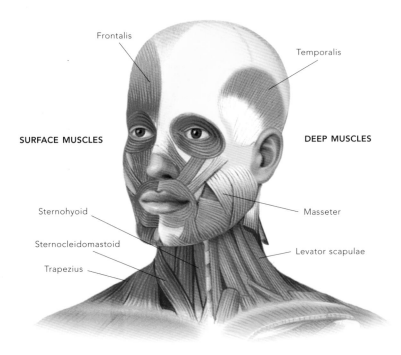

Frontalis

Temporalis

Sternohyoid

Masseter

Sternocleidomastoid

Levator scapulae

Trapezius

Cross section of the neck

THE CROSS SECTION OF THE NECK
REVEALS IMPORTANT STRUCTURES,
SUCH AS THE VOCAL CORDS AND THE
ESOPHAGUS, which are found in the
anterior portion of the neck. On the
posterior side is the spinal cord, and
emanating from that is the brachial
plexus, which supplies nerves to the
upper limb. If this plexus is
compromised, symptoms such as pins
and needles—or numbness—may be
experienced locally or in the arm on
the affected side.

Other structures to consider are
glands—such as the thyroid—and vital
blood vessels—such as the carotid
artery and the jugular vein. The carotid
artery transports blood to the brain,
and the jugular vein transports blood
away from the brain. Any compromise
in this blood flow to the brain can lead
to a higher risk of stroke.

Therefore, it is evident that the neck
plays a vital role in function and in the
transmission of essential nutrients to
the head and brain. Several vital
structures exist within the neck, and
special care must be taken when
looking at or assessing the region.

▶ **STRUCTURES OF THE NECK**
The neck is a complex arrangement of
vital structures, including the carotid
artery and the jugular vein.

CROSS SECTION OF THE NECK

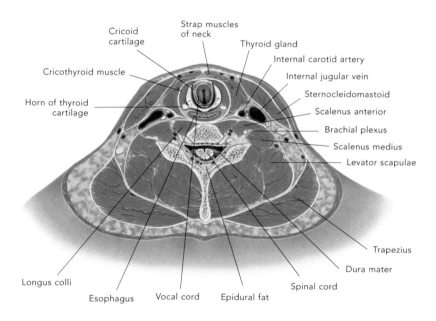

Cricoid cartilage

Strap muscles of neck

Thyroid gland

Internal carotid artery

Cricothyroid muscle

Internal jugular vein

Horn of thyroid cartilage

Sternocleidomastoid

Scalenus anterior

Brachial plexus

Scalenus medius

Levator scapulae

Trapezius

Dura mater

Longus colli

Esophagus

Vocal cord

Epidural fat

Spinal cord

Concussion in sports

Concussion, or mild traumatic brain injury (mTBI), occurs when contact and acceleration forces propagate through the brain. During impact, these forces are transferred to the brain and vascular tissue, causing micro-scale deformations. Traumatic brain injury is classified as mild if the individual experiences loss of consciousness and/or disorientation for less than 30 minutes. Although it is classified as mild, symptoms can be traumatic and include behavioral and emotional disturbances that can be evident post-concussion. More recently, the public has been made aware that repeated mTBI is also a risk factor for a variety of dementia-related neurological dysfunctions, most notably chronic traumatic encephalopathy (CTE). CTE is a progressive neurodegenerative disease that affects people who have received repeated blows to the head. There have been a number of high-profile cases involving famous contact sport athletes, as well as reports of military veterans succumbing to the devastating effects of CTE.

Any athlete who suffers loss of consciousness should be considered as concussed until a full medical assessment has been undertaken. Due to the potential delayed onset of symptoms, individuals who have sustained a suspected concussion should never be allowed to return to the field of play on the same day. All athletes who have sustained a suspected concussion should

▶ **CONTACT SPORTS**

Due to large impact forces to the head and neck, concussion can be common in sports such as football, causing possible damage to the brain.

Concussion in sports *(cont.)*

follow a clearly defined program to return to their activity. Within this process, individuals are carefully monitored as they complete progressive increases in the amount of activity they undertake. This process ranges from a period of full physical and cognitive rest to a return to full-contact training followed by a full return to play.

The protective clothing used in football protects the players from direct impact injuries, but does nothing to stop the violent shaking of the brain caused by the forces received. In the absence of appropriate protective equipment, new technologies are being developed that allow individuals who support athletes to monitor and record the frequency and intensity of forces received by the body. In time, it is hoped that managing the impact forces received by players may lead to a reduction in the long-term side effects of sports-related concussions.

▶ **NECK INJURY**
Hyperflexion of the neck, known as whiplash, is an injury often sustained during automobile accidents.

HYPEREXTENSION

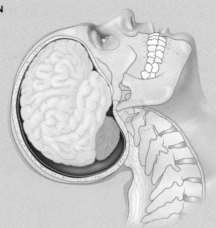

HYPERFLEXION

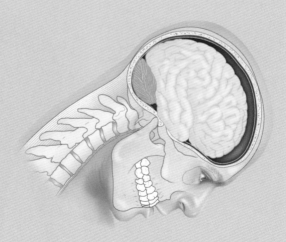

Chapter 13:
Thoracic spine and ribs

The centrally located thoracic spine and ribs are at the core of the human body, providing crucial postural support. As well as protecting the heart and lungs, the ribs play an essential role in breathing, since their rise and fall, along with the relaxation and contraction of the diaphragm, creates the pressure changes that draw air into and out of the lungs.

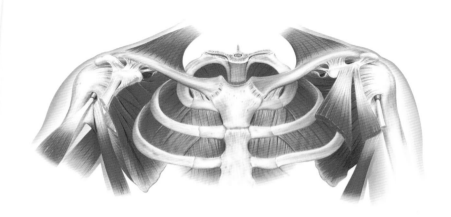

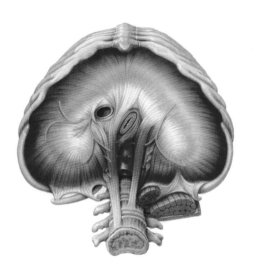

Thoracic spine

THE THORACIC REGION OF THE SPINE IS LOCATED BETWEEN THE CERVICAL AND LUMBAR REGIONS OF THE SPINE. This region contains twelve vertebrae, of which seven are typical (T2–T8) and five are atypical (T1, T9–T12). Typical thoracic vertebrae have heart-shaped vertebral bodies with a small circular foramen, whereas atypical ones do not. Thoracic vertebrae are also characterized by two demi-facets (superior and inferior) on both sides of the vertebral body, which are designed for articulation with the rib heads.

The vertebral bodies of the thoracic spine increase in size as they descend toward the lumbar region. The spinous processes of the thoracic vertebrae project inferiorly, with their tips covering the body of the vertebra adjacent below. Inferior articular processes project inferiorly from the laminae, on which the anterior inferomedially facing articular facets are found. The superior articular processes face superiorly and posteriorly and have oval-shaped articular facets. The transverse processes project posteriorly and

laterally and also contain oval-shaped facets for articulation with the tubercles of the ribs. The superior demi-facet, intervertebral disc, and inferior demi-facet form an articulation for the rib head. The superior demi-facet receives the head of its corresponding rib. For example, the superior demi-facet of T6 articulates with the inferior facet on the head of the sixth rib. The inferior demi-facet will articulate with the superior facet on the head of the seventh rib.

▶ **THORACIC VERTEBRAE**

A typical thoracic vertebra has a heart-shaped vertebral body with small vertebral foramen. On each side of the vertebra are two costal demifacets for articulation with the ribs, a characteristic that is unique to the thoracic spine. The attachments of the ribs increase stability at the thoracic spine but limit the amount of movement available.

THORACIC VERTEBRA

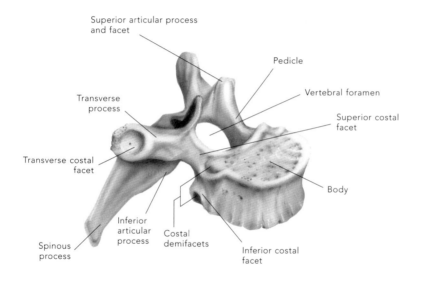

Superior articular process and facet

Pedicle

Vertebral foramen

Transverse process

Superior costal facet

Transverse costal facet

Body

Inferior articular process

Costal demifacets

Spinous process

Inferior costal facet

Thoracic spine *(cont.)*

Atypical thoracic vertebrae share many characteristics with typical thoracic vertebrae but do have some important morphological differences. The first thoracic vertebra (T1) has a long, thick spinous process that projects almost horizontally. The first and second ribs articulate with T1 at its superior costal facet and its inferior demi-facet, respectively. The thoracic vertebrae increase in size and width from T1 to T12. Toward T12, the vertebrae become more supportive and less movable—characteristics of the lumbar spine. This is most apparent at T12, which has a large vertebral body and is without costal facets on its inferior portion. Vertebrae T11 and T12 have triangular spinous processes that project posteriorly but less inferiorly than the other thoracic vertebrae. T12 also displays mammillary processes—small tubercles—that serve as an attachment point for the multifidus muscle. The T12 vertebra is the most commonly injured vertebra due to its exposure to increased levels of stress.

▶ **THORACIC SPINE**

The thoracic spine contains 12 vertebrae and is located between the cervical and lumbar spine. Combined with the sternum and ribs, it forms part of the thoracic cage, vital for protecting the internal organs of the abdomen and thorax. The vertebral bodies of the thoracic spine increase in size as they descend the vertebral canal, becoming more stable but less mobile. The thoracic region has a convex curve, also known as a kyphotic curvature.

THORACIC SPINE REGION

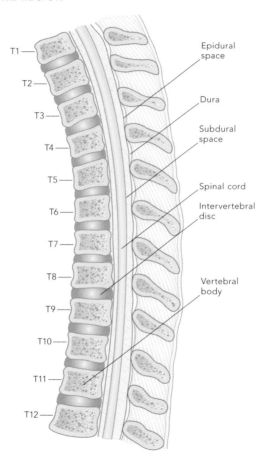

T1

T2

T3

T4

T5

T6

T7

T8

T9

T10

T11

T12

Epidural space

Dura

Subdural space

Spinal cord

Intervertebral disc

Vertebral body

Ribs

RIBS ARE CURVED, FLAT BONES THAT IN CONJUNCTION WITH THE STERNUM AND THE THORACIC VERTEBRAE FORM THE THORACIC CAGE (RIB CAGE). The thoracic cage has three important functions: protecting the organs of the abdomen and thorax, supporting the upper limb, and providing muscular attachment to aid respiration.

Ribs can be characterized as either typical or atypical. Typical ribs are present from the third to ninth ribs and have a head, neck, tubercle, angle, and shaft. The head contains two facets for articulation with corresponding facets on the thoracic vertebrae (superior and inferior facets). The neck of the rib is directed anterosuperiorly and connects the head of the rib with the body. A tubercle is present where the neck meets the body. This is the point of articulation with the transverse process. The body (shaft) of the rib is curved and is most observable at the costal angle, where it turns anterolaterally. The rib body has a convex internal surface and a concave external surface. The convex internal surface is where the costal groove is located, which gives protection to the intercostal nerves and vessels.

The first, second, tenth, eleventh, and twelfth ribs are atypical. The first rib is the shortest and is broad and oriented obliquely from posterior to anterior. The head has one circular facet for articulation with T1. Its superior border is marked by two grooves for the subclavian vein and artery. The medial border is grooved and contains the scalene tubercle at its midpoint, which is the site of attachment for the anterior scalene.

▶ **RIBS**

The ribs protect the vital organs of the chest cavity, with each pair of ribs attached to a thoracic vertebra.

RIBS

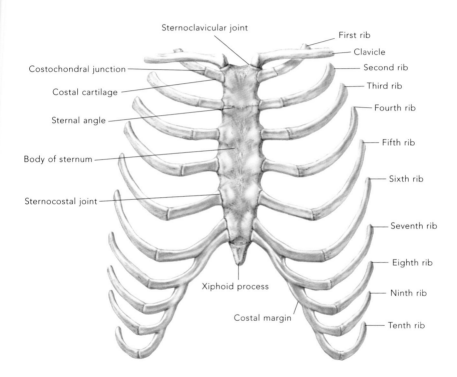

Sternoclavicular joint

First rib

Clavicle

Second rib

Costochondral junction

Third rib

Costal cartilage

Fourth rib

Sternal angle

Fifth rib

Body of sternum

Sixth rib

Sternocostal joint

Seventh rib

Eighth rib

Xiphoid process

Ninth rib

Costal margin

Tenth rib

Pectoral girdle

THE PECTORAL GIRDLE CONSISTS OF
BOTH LEFT AND RIGHT CLAVICLES AND
SCAPULAE, AS WELL AS THE STERNUM.
It connects the bones of the upper
limbs (superior appendicular skeleton)
to the axial skeleton. The pectoral
girdle, unlike the pelvic girdle, does
not attach posteriorly to the trunk.
Therefore, it is often described as an
incomplete ring, allowing each limb
to move independently.

Scapula

The scapula forms the posterior
portion of the pectoral girdle. The
scapula is a flat, triangular bone that
lies obliquely over the posterior portion
of the rib cage (ribs 2–7). The scapula
has two surfaces, a costal and a dorsal
surface. The costal surface is concave,
forming the large subscapular fossa,
while the dorsal surface is separated
into a small and a large section by the
spine of the scapula—the supraspinous
and infraspinous fossae, respectively.
The spine of the scapula is triangular
and projects horizontally from the
dorsal surface, while its lateral aspect
gives rise to the acromion process.

The scapula has a medial, lateral, and
superior border and an inferior,
superior, and lateral angle. The thin
medial border passes between the
superior and inferior angles, whereas
the thicker lateral border passes
between the head (glenoid) and
inferior angle. The superior border is
the shortest of the three and runs
between the superior angle and the
suprascapular notch, medial to the
coracoid process.

Sternum

The sternum is a flat bone that forms
the anterior portion of the thoracic
cage. It offers protection to the vital
organs of the thoracic region, most
notably the heart, and provides
attachment for the costal cartilage.
The sternum is separated into three
parts, the manubrium, body, and
xiphoid process. The manubrium is the
widest, thickest, and most superior part
of the sternum. It can be found around
the level of T3–T4 and articulates with
the clavicle (sternoclavicular joint) and
the costal cartilage of the first and
second ribs. The body of the sternum is

longer and thinner, and it is located at the level of T5–T9. It contains circular facets for articulation with the third to sixth (and a portion of the seventh) costal cartilages. The xiphoid process is the smallest, most inferior part of the sternum, located at the level of T10.

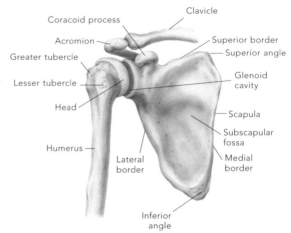

MUSCLES OF THE
PECTORAL GIRDLE

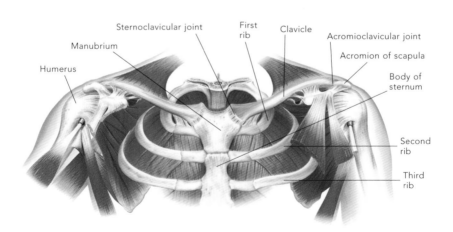

Collarbone

THE CLAVICLE IS MORE COMMONLY KNOWN AS THE COLLARBONE AND CONNECTS THE UPPER LIMB WITH THE TRUNK. It extends horizontally from the sternum (manubrium) to the scapula (acromion) and is S-shaped, with an anteriorly convex border medially (sternal end) and an anteriorly concave border laterally.

The clavicle articulates with the sternum medially to form the sternoclavicular (SC) joint. The articulating surface of the clavicle is larger than the receiving surface on the sternum; there is a disc that divides the joint into a medial and a lateral compartment, each lined with a synovial membrane. Laterally, its small oval facet articulates with the medial aspect of the acromion to form the acromioclavicular (AC) joint. The AC joint is classified as a synovial plane joint.

The clavicle has three functions. First, it provides a rigid point on which the scapula can move freely, keeping it away from the thoracic cage. Second, it can absorb and transmit shock from the upper limbs to the axial skeleton.

Finally, the clavicle provides a protective border (one border of the cervico-axillary canal) for the neurovascular bundles to pass to the upper limb.

The clavicle has a smooth and a rough surface, superiorly and inferiorly, respectively. The inferior surface is rough due to its ligamentous attachments. Strong ligaments bind with the first rib medially and with the scapula laterally. The conoid tubercle is located at the posterior of the lateral inferior border of the clavicle, giving attachment to the conoid ligament.

▶ **THE CLAVICLE**
The clavicle stabilizes the shoulder joint while still permitting a wide range of movement.

COLLARBONE

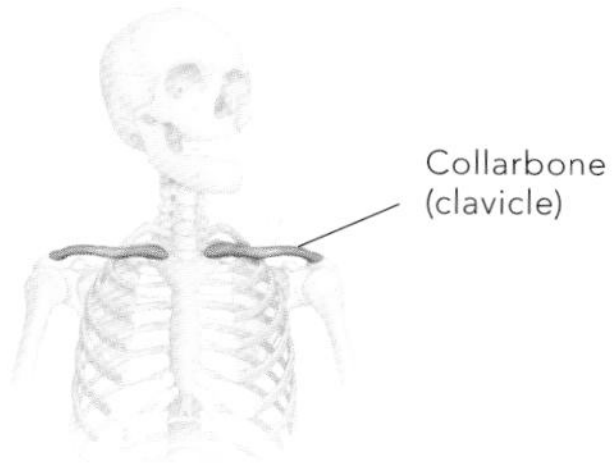

Collarbone
(clavicle)

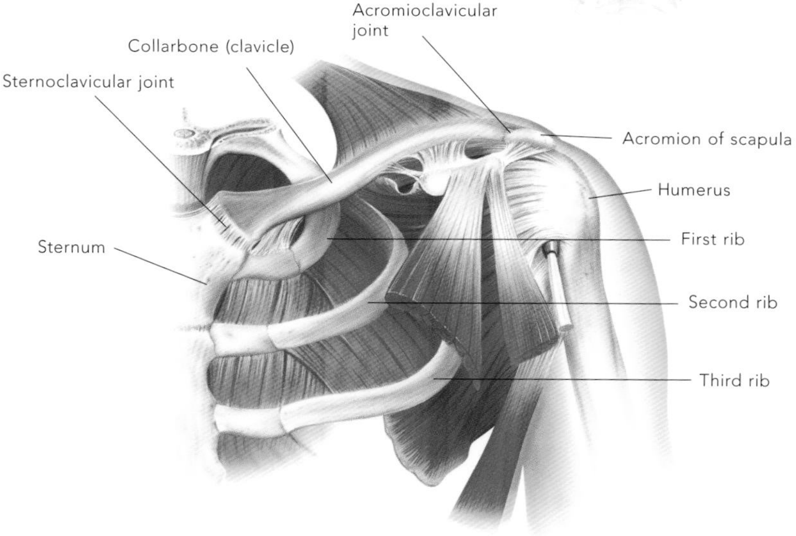

Acromioclavicular
joint

Collarbone (clavicle)

Sternoclavicular joint

Acromion of scapula

Humerus

Sternum

First rib

Second rib

Third rib

Thoracic spinal muscles

THE MOST SUPERFICIAL MUSCLE OF THE POSTERIOR THORACIC REGION IS THE LARGE TRAPEZIUS MUSCLE. This muscular unit's name was derived from its shape. Although technically a bilaterally paired group of muscles, when considered as one unit it resembles a trapezium. The muscle attaches superiorly to the external occipital protuberance, spanning laterally to the posterior acromion and the spine of the scapula. It then descends inferiorly, attaching to the spinous processes of each thoracic vertebra. This large surface area means the trapezius spans most of the upper back, attaching the thoracic spine, shoulder girdle, and base of the skull. The fibers of the trapezius are oriented into three different planes, with each section performing a slightly different functional role. The upper fibers run in a descending fashion from the base of the skull to the lateral side of the scapula, attaching centrally to the bottom cervical and top thoracic spinous processes. The line of pull of this section of the muscle elevates the scapula and shoulder girdle. The middle fibers run near-horizontally from the midthoracic spinous processes to the lateral spine of the scapula and posterior acromion. The action of this segment is retraction of the scapula, pulling the shoulder girdle posteriorly. The lower fibers run in a superior–lateral orientation between the lower thoracic spinous processes and the inferior medial aspect of the spine of the scapula. When contracting, these muscle fibers cause the scapula to depress, lowering the shoulder girdle.

▶ **MOVEMENT OF THE SPINE**
Many of the muscles of the back are involved in providing movement to the spinal region.

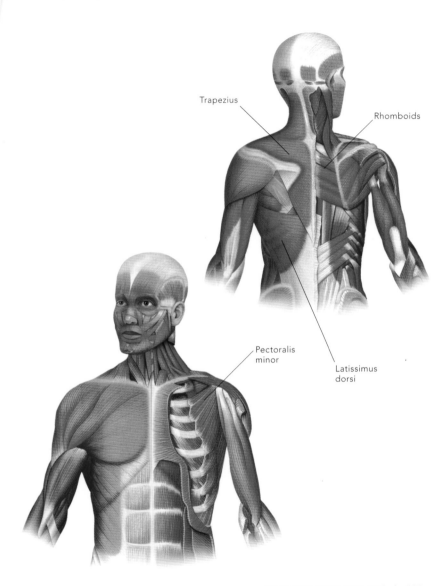

Trapezius

Rhomboids

Pectoralis
minor

Latissimus
dorsi

Thoracic spinal muscles *(cont.)*

The other superficial muscle of the thoracic region is the latissimus dorsi. This large muscle rises from the thoracolumbar fascia of the lower back and the spinous processes of the entire lumbar region and bottom five thoracic vertebrae. Its superior attachment is the intertubercular groove of the humerus. Its action when contracting concentrically is to adduct, extend, and internally rotate the shoulder. Lying deep to the superficial posterior thoracic muscles are the rhomboid major and minor. These short, horizontally oriented muscles run from the spinous processes of C7–T5 and the medial border of the scapula. Their action is to adduct the scapula, stabilizing it against the back of the thoracic cage. The serratus anterior further stabilizes the scapula by protracting and upwardly rotating it. This muscle originates from the outer surface of the upper eighth or ninth rib and runs posteriorly deep to the scapula, attaching to the costal margin of its medial border.

▶ **THORACIC VERTEBRAE**

The thoracic vertebrae offer attachment sites for a variety of muscles and the ribs. They typically have heart-shaped vertebral bodies that ascend in size toward the lumbar region. The spinous process of the thoracic vertebrae project inferiorly, while the transverse processes project posteriorly and laterally. The attachment of the ribs in this segment of the spine increases stability of the region, but also limits the overall range of motion.

THORACIC VERTEBRA AND DISC

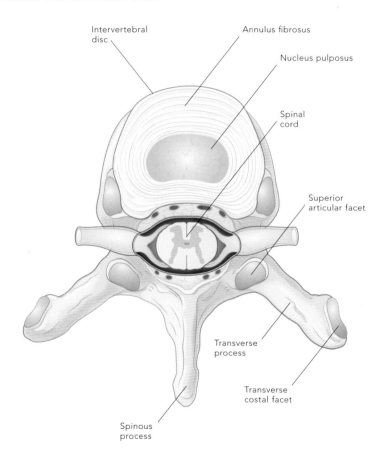

Intervertebral disc

Annulus fibrosus

Nucleus pulposus

Spinal cord

Superior articular facet

Transverse process

Transverse costal facet

Spinous process

Muscles of the thoracic wall

THE MUSCLES OF THE THORACIC WALL INCLUDE THE SERRATUS POSTERIOR, LEVATORES COSTARUM, AND SUBCOSTAL AND INTERCOSTAL MUSCLES. The primary role of this group is to provide stability to the thoracic cage and assist in breathing. The intercostal muscles run from the inferior border of one rib to the superior border of the rib below. The muscles are arranged in three functional levels termed the external, internal, and innermost intercostal layers. All three layers receive innervation from the intercostal nerve. The external intercostal layer helps to elevate the ribs during inspiration; contraction of the other two layers facilitates forced exhalation by depressing the ribs. The subcostal muscles run from the internal surface of the lower ribs, attaching to the superior borders of the ribs two or three layers below. Their action is similar to that of the internal and innermost intercostal groups. The serratus posterior has two separate elements. The superior section runs horizontally from the spinous processes of C7–T3 to the superior borders of the second, third, and fourth ribs. Its contraction elevates the ribs, increasing the diameter of the thoracic cage and forcing inspiration. The inferior section of the serratus posterior runs from the spinous processes of T7–T11 to the inferior border of the eighth to twelfth ribs. Its contraction causes depression of the rib cage, resulting in forced exhalation. The levatores costarum are twelve short fan-shaped muscles, which descend bilaterally from the transverse processes of the bottom six thoracic vertebrae to the adjacent ribs. Their role is stabilization of the rib cage and assisting in its elevation, facilitating inspiration.

▶ **INTER- AND SUBCOSTALS**
The intercostal and subcostal muscles provide stability and flexibility to the thoracic cage.

MUSCLES OF THE THORACIC WALL

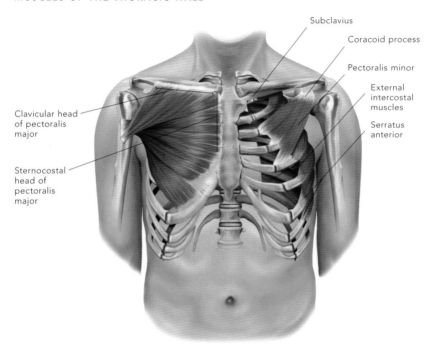

Subclavius

Coracoid process

Pectoralis minor

External intercostal muscles

Serratus anterior

Clavicular head of pectoralis major

Sternocostal head of pectoralis major

The pectoral muscles

THE PECTORAL GROUP CONSISTS OF TWO MUSCLES LOCATED SUPERFICIALLY IN THE ANTERIOR THORACIC REGION. The larger of the muscles is pectoralis major; it has a large surface area and controls various movements of the glenohumeral joint. This thick, fan-shaped muscle is constructed of two separate heads, which blend to share one tendon insertion at the lateral lip of the bicipital groove. The sternocostal head gives rise to the larger section of the muscle, originating from the anterior surface of the sternum, the costal cartilage of the first six ribs, and the aponeurosis of the external oblique muscle. This portion of the muscle's fibers run horizontally from the sternum and vertically from the external oblique attachment. The clavicular head originates from the medial half of the inferior border of the clavicle. Its fibers run obliquely toward the distal tendon insertion at the humerus. When both sections of the muscle contract in unison, adduction and medial rotation of the shoulder occurs. The line of pull of the clavicular allows flexion to occur when this section of the muscle contracts independently. When in the flexed position, an independent contraction of the sternocostal section assists with extending the shoulder toward anatomical neutral.

Lying deep to pectoralis major is the much smaller pectoralis minor. This muscle's superior attachment is to the medial border and superior surface of the coracoid process of the scapula. This short, triangular muscle attaches to the costal cartilage of ribs 3-5 and has a stabilizing effect on the scapula, drawing it inferiorly and causing it to anteriorly tilt toward the rib cage.

▶ **MOVEMENT OF THE SHOULDERS**
The chief role of the pectoral muscles is to provide movement to shoulders.

THE PECTORAL MUSCLES

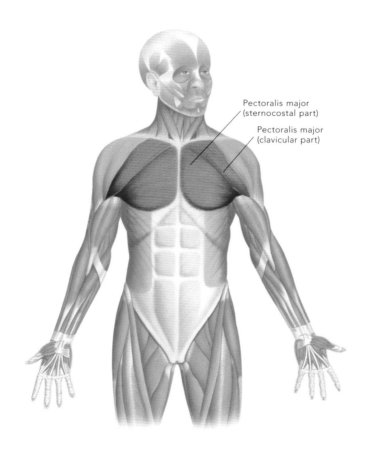

Pectoralis major
(sternocostal part)

Pectoralis major
(clavicular part)

The diaphragm

THE DIAPHRAGM IS A BROAD MUSCULOTENDINOUS STRUCTURE THAT ATTACHES HORIZONTALLY TO THE BOTTOM OF THE RIB CAGE. Its function is essential in controlling the process of respiration. Anatomically, it also serves a purpose in segregating the thorax from the abdominal cavity. Its structure is shaped in a double dome fashion, with its convex superior surface tapering upward toward the thoracic cavity. The right-sided dome sits slightly higher than the left, allowing additional room for the liver, which lies inferiorly. Centrally, the muscle has a thick tendon, which blends anteriorly into the base of the sternum at the xiphoid process. The pericardium of the heart rests on top of this central portion. The main muscular areas of the diaphragm are located toward the outside of the structure. The muscle fibers run horizontally from attachment sites around the base of the rib cage and lumbar vertebrae, toward the central aponeurosis and tendon, so that as the contractile sections shorten, the central area becomes taut and is drawn inferiorly. This process initiates the inspiration phase of the respiratory cycle. Since the diaphragm sits at the top of the abdominal cavity, it plays an essential role in regulating intra-abdominal pressure. This function is an important factor in maintaining the stability of the lumbar region of the spine.

▶ **THE DIAPHRAGM**

Several structures pass through the diaphragm, inculding the major structures of the esophagus, the aorta, and the inferior vena cava.

THE DIAPHRAGM

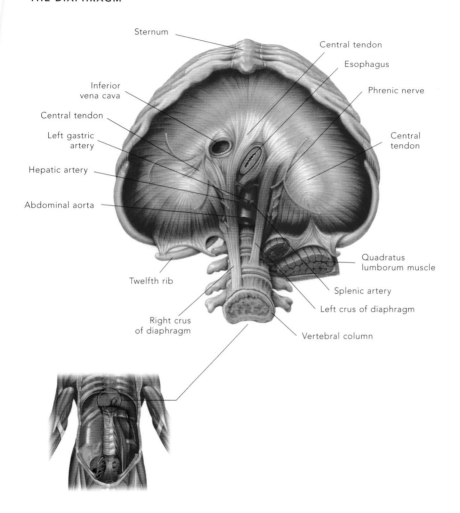

Sternum

Central tendon

Esophagus

Inferior vena cava

Phrenic nerve

Central tendon

Central tendon

Left gastric artery

Hepatic artery

Abdominal aorta

Twelfth rib

Quadratus lumborum muscle

Splenic artery

Left crus of diaphragm

Right crus of diaphragm

Vertebral column

The mechanics of breathing

BOYLE'S LAW DESCRIBES THE RELATIONSHIP BETWEEN THE PRESSURE AND THE VOLUME OF A GAS. The law states that if volume increases, the pressure must decrease. Changes in the pressure of the thoracic cavity draw air in and out of the lungs. As intrathoracic pressure drops below that of the atmosphere, air is drawn into the lungs, beginning the process of respiration and oxygen diffusion into the bloodstream. The respiration process is triggered by contractions of a number of muscles that attach to the thoracic cage. As the muscular outer section of the diaphragm contracts, it draws the central part of its structure inferiorly. As this action occurs, a subsequent co-contraction of the external layer of the intercostal muscles, superior serratus posterior, and levatores costarum occurs. This process of coordinated muscle contraction elevates and expands the rib cage. The increase in volume and resultant decrease in pressure of the rib cage triggers the process of inspiration, drawing air into the lungs.

Exhalation, or forcing air out of the lungs, occurs as the diaphragm and the breathing muscles relax. This process depresses the rib cage and decreases its volume. The pressure inside the rib cage then increases to more than that of the external atmosphere, making the air flow out of the lungs. In general, the process of exhalation is a passive one. However, in some cases, this action can become more forceful. For example, as an individual forcibly blows out to extinguish a candle, the flow of air is stronger than that caused solely by a change in pressure. In this scenario, a process of forced exhalation occurs.

To achieve this action, additional muscles, such as the internal and innermost intercostal muscles and the inferior portion of serratus posterior, are recruited. Muscle contraction further depresses and decreases the size of the rib cage, and the resultant increase in pressure further forces air from the lungs.

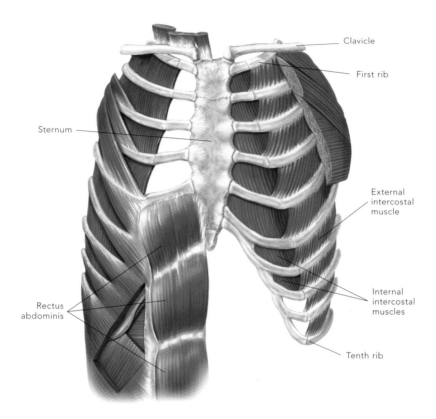

Clavicle

First rib

Sternum

External
intercostal
muscle

Internal
intercostal
muscles

Rectus
abdominis

Tenth rib

▲ **THE INTERCOSTAL MUSCLES**

The external intercostal muscles lie
over the internal intercostal muscles,
and collectively they extend between
the upper and lower borders of
neighboring ribs.

Chapter 14:
Lumbar spine and pelvis

This area of the body combines stability and strength with flexibility and movement. Powerful muscles help to deliver the core stability that is essential for many movements, with strong bones ensuring that the body is capable of withstanding the forces that are produced and received during fast and powerful movements.

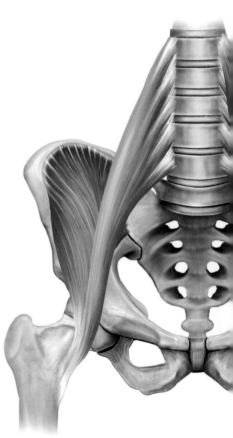

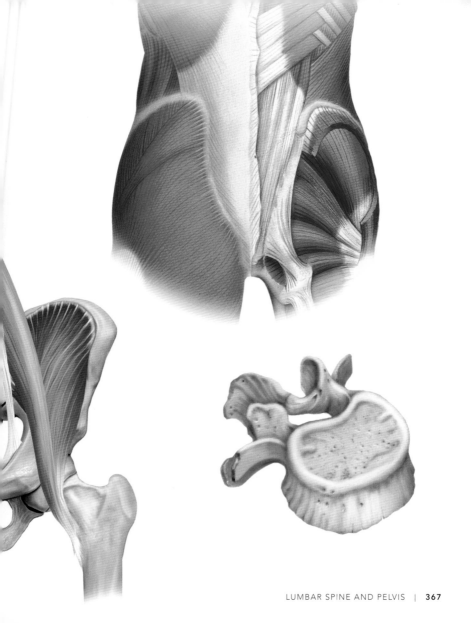

Lumbar vertebrae

THE LUMBAR REGION OF THE SPINE CONNECTS THE THORACIC REGION AND THE SACRUM. It consists of five large, irregular vertebral bones. As with the other regions of the spine, these five bones are identified by number, from superior to inferior according to their position (L1–L5). These vertebrae have large central bodies, allowing for even force distribution and providing a suitable base for supporting the weight of the head and more superior regions of the spine. Since the weight and force that these vertebrae must withstand is far greater than those in other regions

of the spine, they are much larger in size. The total mass of the five lumbar vertebrae is roughly twice that of the seven vertebrae that make up the cervical region.

The lumbar vertebrae display a uniform shape and structure, with short, thick laminae and pedicles. Each segment has transverse processes that project almost laterally from the large spinal body. In general, the transverse processes are thin and tapered. However, L5 is atypical of this region, presenting instead with thick, broad transverse processes. The increased size

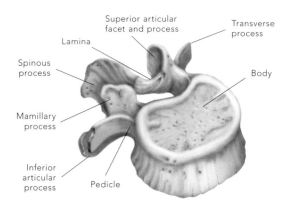

Superior articular
facet and process

Transverse
process

Lamina

Spinous
process

Body

Mamillary
process

Inferior
articular
process

Pedicle

LUMBAR VERTEBRA

of the L5 transverse processes allows for a bilateral attachment site of the iliolumbar ligaments. These ligaments attach each side of the L5 vertebra to the iliac crest of the upper pelvis, limiting movement between these structures while also having a stabilizing effect on the sacroiliac joint. On the posterior surface of each transverse process is an accessory process that acts as an attachment to the small, stabilizing intertransversarii muscles. Each vertebra has a set of superior and inferior articular facets that allow for articulation with the adjacent spinal bone. The posterior surface of the superior articular facets includes a muscular attachment site for the multifidus muscle known as the mamillary process.

The articular facets of the lumbar vertebrae are oriented in a near-vertical fashion. This facilitates flexion and extension in the region while limiting the ability of segmental rotation to occur.

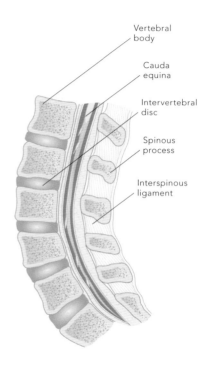

Vertebral body

Cauda equina

Intervertebral disc

Spinous process

Interspinous ligament

LUMBAR SPINAL REGION

The sacroiliac joint

THE SACRUM IS A LARGE, TRIANGULAR BONE LOCATED AT THE BASE OF THE SPINE AND IS FORMED FROM FIVE INDIVIDUAL VERTEBRAE THAT FUSE TOGETHER IN EARLY HUMAN DEVELOPMENT. The lateral border of the sacrum articulates with the ilium bone of the pelvis to form the sacroiliac joint. This articulation is important for the transmission of force between the upper body, trunk, and lower limb. The joint is stabilized by both a highly congruent articulation between the bones, and several thick ligaments. Anterior stability is provided to the joint via the anterior sacroiliac ligament, which forms part of the synovial joint capsule. This is further reinforced by the interosseous sacroiliac ligament and the posterior sacroiliac ligament complex. These broad, thick structures run upward and obliquely from the sacrum to the ilium, meaning that as a downward axial load is

LOCATION OF THE SACROILIAC JOINT

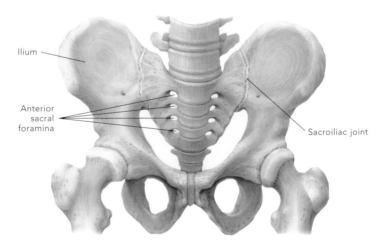

Ilium

Anterior sacral foramina

Sacroiliac joint

applied, the ilia are pulled medially, further stabilizing the joint.

Throughout childhood and early puberty, the sacroiliac joint is freely movable. However, in early adulthood, a process of ossification begins that fuses this joint together. This provides additional stability but limits its movements. The movements that occur at the sacroiliac joint include a small degree of translation between the bones, known as nutation and counter-nutation. These movements are hard to measure and represent minimal displacement. In nutation, the base (top) of the sacrum tilts anteriorly in the sagittal plane against the pelvis, while in counter-nutation the opposite occurs, with the sacrum tilting posteriorly against the ilium.

SACROILIAC JOINT

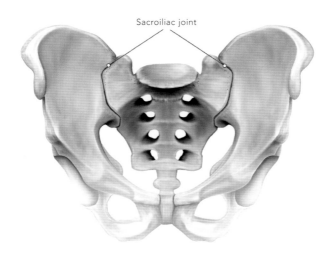

Sacroiliac joint

The pelvis

THE PELVIS IS FORMED BY TWO SETS OF FUSED OSTEOLOGICAL STRUCTURES KNOWN AS THE INNOMINATE BONES. These structures are joined together anteriorly by the cartilaginous symphysis pubis, with posterior articulations to the sacrum via the sacroiliac joint. The pelvis performs three very different, yet equally important, functions. First, the structure provides an attachment point for various muscles of the lower limb

and trunk. Second, the pelvis provides protection to various reproductive, digestive, and genito-urinary organs, which are located in this area of the body. Finally, the pelvis has an important role in transmitting and dissipating force between the body's upper and lower quadrants.

Each innominate is formed of three fused bones: the pubis, ilium, and ischium. When viewed from the exterior, the innominate bones display

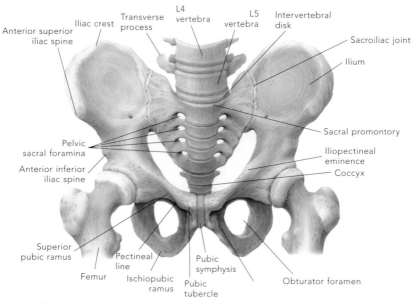

three key features. The superior section of the innominate is formed by the ilium, and the top of this area fans out and forms the prominent iliac crest. Inferiorly, toward the bottom third of the innominate, a point of convergence between all three bones occurs. This deep, cup-shaped feature is called the acetabulum. This structure articulates with the head of the femur to form the hip joint. Just inferior to the acetabulum lies the obturator foramen.

This circular feature is the largest foramen in the body and is covered with a thick collagenous membrane that offers an attachment site for the obturator externus and obturator internus muscles.

▼ **PELVIS**
The male pelvis (left) and the female pelvis (right).

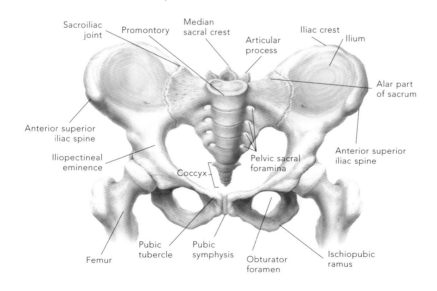

Sacroiliac joint

Promontory

Median sacral crest

Articular process

Iliac crest

Ilium

Alar part of sacrum

Anterior superior iliac spine

Iliopectineal eminence

Coccyx

Pelvic sacral foramina

Anterior superior iliac spine

Femur

Pubic tubercle

Pubic symphysis

Obturator foramen

Ischiopubic ramus

The ilium

THE ILIUM IS A LARGE, FLAT BONE THAT EXTENDS UPWARD FROM THE BOTTOM THIRD OF THE PELVIS. Its superior aspect is a curved edge known as the iliac crest, and this feature is often easily visible in the human body. The most anterior point of the iliac crest forms a bony prominence termed the anterior superior iliac spine (ASIS). This bony point acts as an attachment site for the sartorius muscle and is easily palpable. Just inferior to this point lies a smaller bony protuberance termed the anterior inferior iliac spine (AIIS), which acts as the proximal attachment for the rectus femoris muscle of the quadriceps group. Posteriorly, the iliac crest extends toward another bony edge known as the posterior superior iliac spine (PSIS). When observing the skin, this structure is usually easily visible and appears as a small dimple in the lower back. Inferior to this structure is the less prominent posterior inferior iliac spine (PIIS), which marks the uppermost point of the greater sciatic notch—an opening that allows the passage of the lumbosacral plexus nerve bundle to pass toward the lower limb. On the external surface of the ilium are the faint posterior, anterior, and inferior gluteal lines. These features act as the point of attachment for the gluteus maximus, gluteus medius, and gluteus minimus, respectively. The internal aspect of the ilium presents as a soft, convex curve known as the iliac fossa. This area acts as an attachment site for the hip flexor muscle, the iliacus.

▶ **ASIS**

The anterior superior iliac spine is the point of attachment for the sartorius muscle.

THE ILIUM

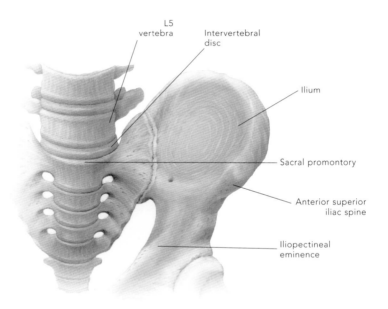

L5
vertebra

Intervertebral
disc

Ilium

Sacral promontory

Anterior superior
iliac spine

Iliopectineal
eminence

The ischium

THE ISCHIUM IS A SMALL, IRREGULAR BONE THAT EXTENDS INFERIORLY AND POSTERIORLY FROM THE ILIUM AT THE POINT OF THE ACETABULUM. The ischium is fused to the ilium across the posterior aspect of the pelvis toward the base of the greater sciatic notch. At this point, a sharp, bony protuberance on the top of the ischium, termed the ischial spine, can be located. This point indicates the beginning of the lesser sciatic notch. The ischium then projects further in an inferior direction to form the large, broad, bony structure of the ischial tuberosity. This site acts as an attachment site for various lower limb muscles, such as the hamstring group.

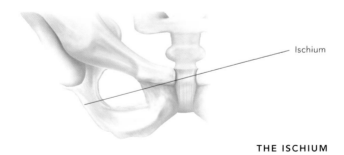

Ischium

THE ISCHIUM

The pubis

THE PUBIS IS THE SMALLEST BONE OF THE INNOMINATE. It descends from the acetabulum in an anteromedial direction via the superior pubic ramus before flattening out to become the body of the pubis. Running between the superior ramus and body of pubis is a distinct feature known as the pectineal line, which acts as an attachment site for the pectineus muscle. The superior border of the body is termed the pubic crest, and its medial aspect is the site of the distal attachment of the rectus abdominis.

The medial ends of the pubis, from both sides of the body, meet in the center to form the pubic symphysis joint. This firm articulation allows minimal movement to occur, with roughly 2 mm of translation and a small amount of rotation available. Joint stability is provided by a tough cartilaginous central disc and the thick fibers of the inferior and superior pubic ligaments.

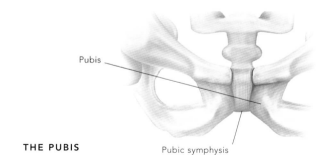

Pubis

THE PUBIS

Pubic symphysis

Muscles of the lower back, abdomen, and pelvic floor

MOST OF THE BODY'S WEIGHT LIES ANTERIOR TO THE SPINAL COLUMN, MEANING THAT THE MUSCLES OF THE BACK FUNCTION TO PROVIDE POSTERIOR STABILITY TO THE SPINE. The muscles of the lower back can be divided into two clear groups: extrinsic and intrinsic muscles. The extrinsic muscles control limb movements and respiratory functions, while the intrinsic muscles aid local stability of the spine, ensuring an upright posture.

MUSCLES OF THE LOWER BACK

SUPERFICIAL   DEEP

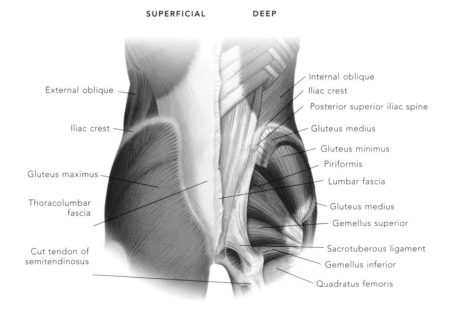

External oblique

Iliac crest

Gluteus maximus

Thoracolumbar fascia

Cut tendon of semitendinosus

Internal oblique

Iliac crest

Posterior superior iliac spine

Gluteus medius

Gluteus minimus

Piriformis

Lumbar fascia

Gluteus medius

Gemellus superior

Sacrotuberous ligament

Gemellus inferior

Quadratus femoris

Extrinsic muscles of the lower back

The latissimus dorsi is a large, fan-shaped muscle that attaches directly to the transverse processes of the bottom five thoracic vertebrae before blending into the thoracolumbar fascia. The latissimus dorsi creates movement at the shoulder joint; however, it may also play a role in stabilizing the lower back by creating a torsion in the thoracolumbar fascia when contracting.

Intrinsic muscles of the lower back

The erector spinae muscle group sits between the spinous process and transverse process of each vertebra from the lumbar region upward to the base of the skull. The erector spinae muscles are the primary extensors of the spine and play a key role in the maintenance of an upright posture. This group is subdivided into three columns; the spinalis forms the most medial column, with the longissimus and iliocostalis lying more laterally. Beneath this is the transversospinalis group of muscles consisting of the semispinalis, multifidus, and rotatores. These muscles sit in the groove formed between the spinous and transverse processes of the vertebrae.

Abdominal muscles

The anterior abdominal wall consists of four bilaterally paired sets of muscles. Three of these pairs—the external obliques, internal obliques, and transverse abdominis—are flat in nature, with horizontally running muscle fibers. The final set, the rectus abdominis, runs vertically and has a much larger cross-sectional area. The rectus abdominis originates from the superior aspect of the pubic crest and inserts into the base of the ribs at the xiphoid process. The orientation of this muscle's fibers means that its primary action—under concentric contraction—is forward flexion of the spine. Functionally, the role of the rectus abdominis may also be to contract both isometrically and eccentrically to control spinal extension. The deepest-lying muscle of the anterior abdominal wall is the transversus abdominis, which extends from the iliac crest and the bottom six ribs. The muscle fibers' horizontal orientation does not allow it to generate torque or any subsequent movement of the spine. The contraction of this muscle is important because it compresses the abdominal cavity and creates torsion that aids stability in the lumbar region. Lying superficial to the transverse abdominis are the internal obliques and the external obliques. These muscles also have muscle fibers arranged in a horizontal fashion; however, the movements of flexion of the trunk and posterior pelvic tilt can be attributed to their contraction.

▶ **THE ABDOMEN**

The major muscles of the abdomen include rectus abdominis and the external and internal oblique muscles.

DEEP MUSCLES OF THE ABDOMEN

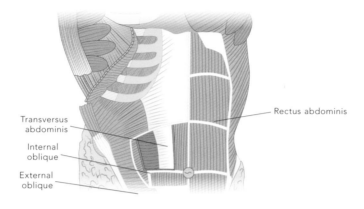

Transversus
abdominis

Internal
oblique

External
oblique

Rectus abdominis

SURFACE MUSCLES OF THE ABDOMEN

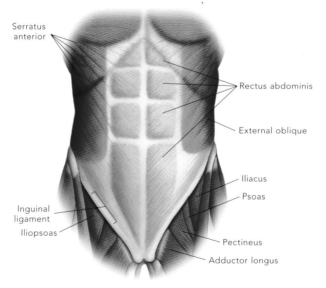

Serratus
anterior

Rectus abdominis

External oblique

Iliacus

Psoas

Inguinal
ligament

Iliopsoas

Pectineus

Adductor longus

Abdominal muscles *(cont.)*

The posterior abdominal wall is constructed of two large torque-generating muscles: the quadratus lumborum and the psoas major. The quadratus lumborum lies just anterior and attaches to each of the transverse processes of the top four lumbar vertebrae. Its distal attachment site is the anterior lip of the medial iliac crest. Its proximity to the spine provides stability to the lumbar region when a bilateral contraction is initiated. Under unilateral contraction, lateral flexion of the spine occurs. The psoas major lies deep to the quadratus lumborum and rises from the transverse processes of T12–L5 and the lateral aspects of the discs between these vertebrae. The muscle blends with the iliacus to share a common tendon that inserts into the lesser trochanter of the femur. The primary role of this muscle is to flex the hip, and its secondary role is in lateral flexion and vertical stabilization of the lumbar spine.

▶ **THE ABDOMINAL WALL**

The muscles of the abdominal wall provide support to the internal organs and play a role in breathing.

POSTERIOR ABDOMINAL WALL

Psoas major

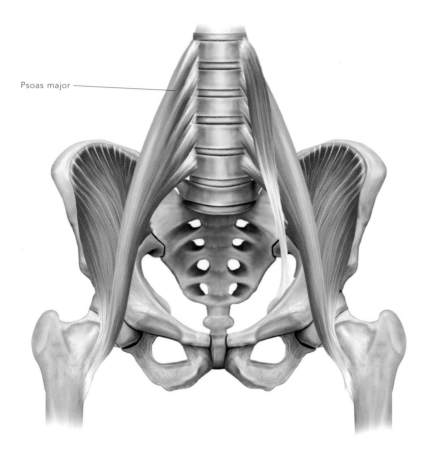

The pelvic floor

THE PELVIC FLOOR MUSCLES FORM THE BASE OF THE ABDOMINOPELVIC CAVITY. They run horizontally and descend from the bony elements of the inferior pelvis. Within this area, the deep, bowl-shaped pelvic diaphragm is constructed from the fibers of the coccygeus and levator ani muscles. The pelvic floor is surrounded by a thick layer of connective tissue.

The coccygeus descends from the ischial spine and inserts into the inferior end of the sacrum and coccyx. The larger levator ani muscle descends posteriorly from the pubic body, blending into the overlying fascia of the perineum and attaching posteriorly to the inferior sacrum via the sacrococcygeal ligament. Visually, the pelvic diaphragm appears like a hammock suspended from the lower pelvic bones. Toward the anterior portion is an opening termed the urogenital hiatus, which allows passage for the urethra (and in females the vagina). The function of the pelvic floor muscles is to provide support for the genitourinary organs, which lie within this area. These muscles also play a regulatory function in the maintenance of inter-abdominal pressure and as part of the urinary and anal sphincters that control continence.

▶ **THE PELVIC FLOOR**

The pelvic floor muscles straddle the pelvis, forming a muscular cradle for the pelvic organs.

PELVIC FLOOR MUSCLES

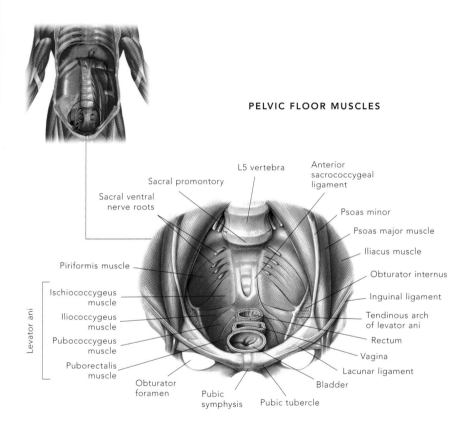

Sacral promontory

L5 vertebra

Anterior sacrococcygeal ligament

Sacral ventral nerve roots

Psoas minor

Psoas major muscle

Iliacus muscle

Piriformis muscle

Obturator internus

Ischiococcygeus muscle

Inguinal ligament

Iliococcygeus muscle

Tendinous arch of levator ani

Pubococcygeus muscle

Rectum

Puborectalis muscle

Vagina

Lacunar ligament

Obturator foramen

Bladder

Pubic symphysis

Pubic tubercle

Levator ani

Core stability in sports

The muscles of the lower back, abdominal wall, and pelvic floor are essential for providing stability to the lumbar region of the spine. These muscle groups are often referred to together as the core. Within sports, this region of the body experiences high levels of force that challenge these muscles to perform at the highest level. As such, proper and effective conditioning of this area is essential. Understanding the functional anatomy of these muscles is important for anyone involved in developing training programs aimed at improving performance and limiting injury to athletes.

The SAID (Specific Adaptation to Imposed Demands) principle suggests that it is important to train muscles for the functional role that they undertake.

Muscles that are close to the spine—such as the multifidus group—use sustained contractions to promote segmental stability and maintain an upright posture. This suggests that these muscles need to be trained for endurance, as opposed to strength training. As such, it is common practice for core exercises to be prescribed to athletes that require them to hold static postures for prolonged periods of time. The very nature of sporting movements means that they are highly dynamic, which is quite contrary to the sustained static posture prescribed in these exercises. Static holds may offer a basic level of conditioning for local stabilizing muscles of the spine. However, additional strategies must also be developed to condition the muscles for the exact demands of the sport being played.

It may also be the case that the functional role of some muscles differs from that assumed by the basic origin and insertions of their structural anatomy. For instance, the rectus abdominis is often viewed as having a primary function that causes forward flexion of the spine. For this reason, the muscle is often trained via the forward flexion sit-up exercise. This action is not overly functional, and in sporting settings, athletes very rarely need to forward-flex their trunk using their rectus abdominis. In reality, the action of this muscle may be to resist excessive lumbar extension by both isometrically and eccentrically contracting. Training the muscle just to shorten, as in the case of the sit-up exercise, may not effectively condition it to the requirements of the functional tasks it undertakes.

ISOMETRIC EXERCISE —THE PLANK

DYNAMIC EXERCISE —WOODCHOP

▲ **CORE-STRENGTHENING EXERCISE**

Core exercises, such as the two shown above, can provide stability to the lumbar region of the spine.

Chapter 15:
Gait

Humans are unique with their bipedal form of movement, standing vertically on two limbs. As the speed of movement increases, humans change gait from walking to running, with many elite athletes capable of speeds in excess of 25 miles per hour. As the only point of contact with the ground, the feet must produce and withstand the forces that are needed to propel the body in the desired direction of travel.

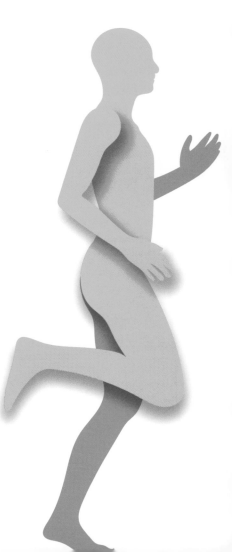

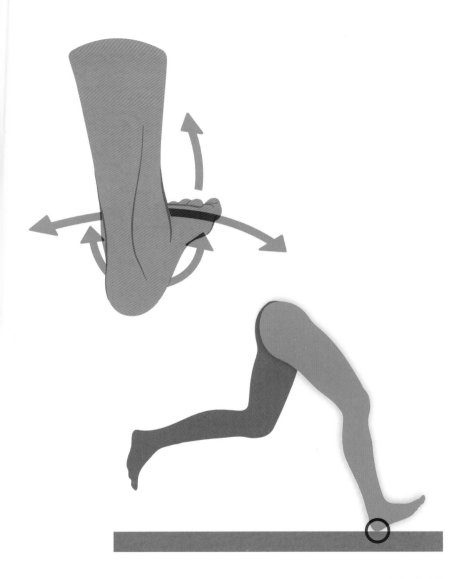

Gait overview: Walking and running

THE TERM "GAIT" REFERS TO THE ACTION OF WALKING OR RUNNING. Gait is important, as it is believed that deviations from normal gait may play a role in musculoskeletal injury, particularly to the lower limb and potentially to the trunk. Knowledge of the different phases of gait and the interactions between the segments of the lower limb is vital to be able to effectively analyze an individual's gait, and therefore to identify the possible contributory factors to the injury an individual may have. There is a difficulty with this because although there is a supposed "normal gait," there are differences between individuals in their size, shape, and mechanics, and so it may be better to consider the characteristics of gait within an ideal range rather than an exact norm. For example, the angle of dorsiflexion at the ankle during the stance phase of gait should be considered within

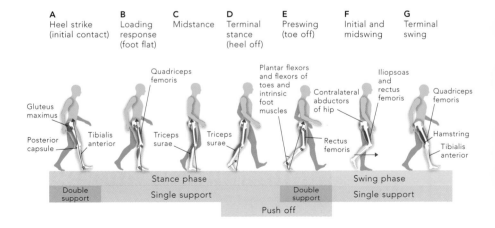

A Heel strike (initial contact)

B Loading response (foot flat)

C Midstance

D Terminal stance (heel off)

E Preswing (toe off)

F Initial and midswing

G Terminal swing

ranges rather than a specific angle. Additionally, an individual may appear to have an abnormal gait but be able to function at a high level and never incur an injury.

The gait cycle can be broken down into different phases linked to whether the limb is in its stance phase or its swing phase. Within the stance and swing phases are further subphases, which can be seen in the image. The majority of research into abnormal or pathological gait has focused on the stance phase and issues with where the initial contact is on the foot and how the loading response occurs. Further attention is also given to the toe-off phase, as this can affect the propulsion of the individual.

◀ **WALKING GAIT CYCLE**

There are seven stages of the walking gait cycle, shown here from A to G, which cover the two principal phases, stance and swing. Within these are the subphases double support (where both feet are on the ground) and single support (where one foot is on the ground). The major muscles at work are also indicated.

Gait overview:
Walking and running *(cont.)*

Running gait differs from walking gait. There are still phases during running, but there is a period where there is no ground contact with either limb. The length of time of this noncontact depends on the speed at which the individual is traveling. There is also likely to be more deviation in foot strike in relation to the point on the foot where it first contacts the ground. This in itself can have significant effects on the mechanics of the rest of the lower limb as the different segments interact with each other. The image shows the external forces acting on a runner, but it is important to note that these forces are also occurring during walking, although some will be reduced. The image farthest right shows one type of pathological gait, and the key issue to note is the consequent effects of this hip drop on the rest of the lower limb due to the close interactions of the other segments.

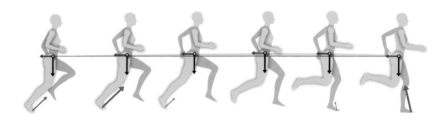

F_d Aerodynamic drag force   ●— Center of mass of runner

F_r Ground reaction force F_g Force of gravity

Force vectors not drawn to scale

NORMAL ALIGNMENT

HIP DROP

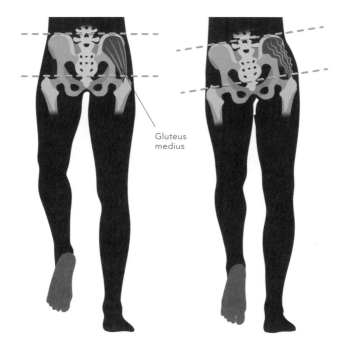

Gluteus medius

◀ **RUNNING GAIT CYCLE**

Illustration of the forces at play during the running gait cyle.

▲ **HIP DROP**

The hip drop shown in the image above, known as the Trendelenburg sign, can have consequences on the movements in the lower limb.

Foot strike in gait

"FOOT STRIKE" REFERS TO THE INITIAL CONTACT THAT THE FOOT MAKES WITH THE GROUND, WHICH CAN VARY BETWEEN INDIVIDUALS. During walking, it is generally expected that initial contact will be made with the heel, and ideally it should be just lateral to the midline of the heel. This may be affected by the position of the rearfoot of an individual—the variations of this can be seen in the image below. The starting position of the rearfoot may cause an individual to favor initial contact in another particular region, which in turn will have an effect on the movement of the foot in the loading response phase.

It may also be apparent that some individuals will make initial ground contact with the midfoot, while in others the initial contact will be the forefoot. If initial contact is on these areas during walking, it is deemed to be abnormal and may be due to a neurological issue or potentially a restriction elsewhere in the body that is

PRONATION AND SUPINATION

Pronated · Neutral · Supinated

THE RUNNING CYCLE

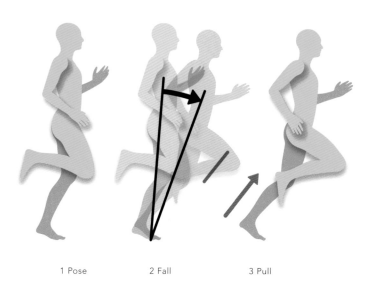

1 Pose   2 Fall   3 Pull

▲ **GROUND CONTACT**

Making initial contact with the ground with the forefoot is seen more commonly in running and changes the mechanics of the gait cycle and the forces acting upon it.

Foot strike in gait *(cont.)*

causing a compensation, or adjustment, in the foot strike.

During running, it is more common to observe an individual with a midfoot or forefoot strike than in walking, although this is often dependent on the distance being run or the speed of the runner. Traditionally, rearfoot running has been the style of running analyzed in scientific literature. Many of the shoe types available are designed to accommodate a rearfoot running style,

and the same principles of walking gait can be applied to runners. A rearfoot strike, followed by pronation, allows for the foot to absorb some of the ground reaction forces and provides time for the foot to resupinate to enable propulsion.

Forefoot running has become very popular in recent years and has been claimed to be the natural way that a human being runs. Shoes have been designed to try and replicate this style of running, and proponents have also

FOOT STRIKE

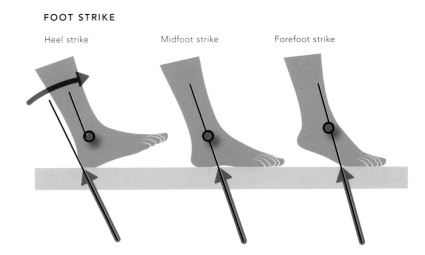

Heel strike   Midfoot strike   Forefoot strike

▼ REARFOOT STRIKE

A rearfoot ground strike can aid in the absorption of the forces created by the impact.

Foot strike in gait *(cont.)*

advocated barefoot running. The image on page 409 shows how the mechanics can change when there is a forefoot strike and demonstrates how the direction of the ground reaction force can change with this type of foot strike. There is a plethora of research suggesting that this approach to running may be beneficial and may reduce the risk of musculoskeletal injury to the lower limb. This can be linked to the mechanics of the foot and the motions that occur during gait when there is a heel strike compared to a forefoot strike. At present, the evidence is inconclusive on the best approach to running, and individuals should instead find the style of running that best suits them in relation to injury prevention and performance.

A further consideration applies to those involved in sports that are multi-directional. As can be seen in the image, lateral motion will have different effects on the mechanics of the body and the ground reaction forces placed upon it. Gait analysis in its simplest form may not be as useful as it seems, if only considering motion in the sagittal plane. The human body moves in three planes, and the majority of movements in both sports and everyday activities take place in all three planes. Gait may appear to be a movement that occurs only in one plane, but it is important to note that this is also a multiplane movement and, as such, there may be issues to be addressed in any of these planes.

▲ **FOOT STRIKE IN SPORTS**

While walking and running gaits should be considered, those playing sports move in different directions, causing different foot strike patterns.

Motion of the foot

FOLLOWING FOOT STRIKE, A COMPLEX SERIES OF MOTIONS OCCUR IN THE FOOT IN ORDER TO ABSORB THE FORCES PLACED UPON THE LIMB. Further motions are required as the limb transfers weight onto the other limb. The image opposite shows a relatively simplistic view of how the foot moves from heel strike to toe off. It gives three examples of how the foot can move through the stance phase. It should be noted that the ideal motion does involve a degree of pronation, and the proposal is that overpronation or underpronation may lead to problems either in the foot or further up the chain.

Following heel strike, there needs to be a degree of pronation to enable the foot to absorb the ground reaction forces. Pronation is a combination of movements at the subtalar joint and the midtarsal joints. Pronation allows force to be distributed across a greater surface area because it transfers the force medially toward the medial longitudinal arch of the foot. The image on page 185 showing the line of the center of pressure through the foot demonstrates

the ideal motion through the stance phase. Following pronation, the foot needs to resupinate, meaning the joints in the foot are locked to create a rigid lever from which the individual can propel themselves forward onto their other limb. It is possible that there may be deviations at any point in the stance phase, depending on any restrictions or weaknesses that an individual may have elsewhere in the body.

A pronated, or everted, rearfoot will result in the center of pressure beginning medial to the desired position and will therefore likely continue throughout the movement. The individual is then more likely to overpronate. In contrast, someone who strikes on the far lateral side of their heel is said to be as underpronating, which describes the foot as remaining in a relatively supinated position. If an individual tends to bear weight on the lateral side of the foot for too long, they can then pronate quickly—or pronate too late—leading to their foot slapping down on the ground. In either instance, the individual is unlikely to have their foot in the ideal

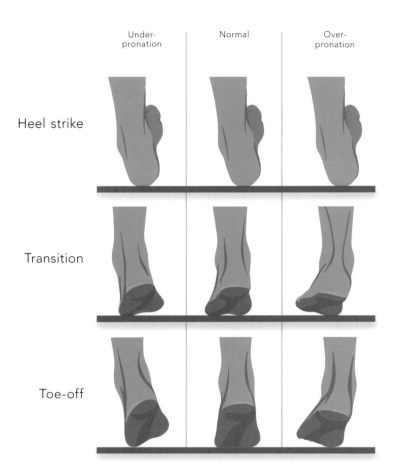

Under-pronation · Normal · Over-pronation

Heel strike

Transition

Toe-off

Motion of the foot *(cont.)*

resupinated position to then enable appropriate propulsion.

The relationship between the motion of the foot and the motion within the lower limb are linked. When pronation occurs, there is then an internal rotation motion of the tibia followed by internal rotation of the femur. When the foot resupinates, the opposite motion occurs—the tibia and femur externally rotate. Problems can occur when this sequence of motions is mistimed or does not occur. The sequence may be mistimed if there are weaknesses in the muscles involved in gait or if the joints in the lower limb are restricted.

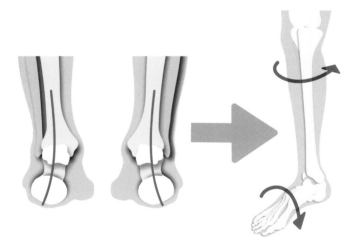

▲ **PRONATION**

There is a link between motion of the foot and of the lower limb. Pronation should be followed by internal rotation of the tibia and femur.

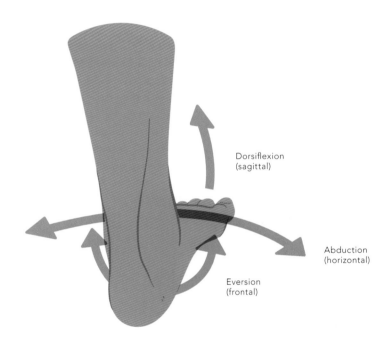

Dorsiflexion
(sagittal)

Abduction
(horizontal)

Eversion
(frontal)

▲ **PRONATION**

It should be noted that pronation
involves motion in all three planes.

Hip and knee angles in gait

BOTH THE HIP AND THE KNEE FLEX EXTEND AT DIFFERENT POINTS IN THE GAIT CYCLE. At initial contact, the hip will be slightly flexed and as weight is accepted and transferred, the hip gradually moves through to an extended position. The degree to which this happens will vary between individuals and can be affected by hip anatomy as well as any muscular restrictions, such as shortened hip flexors that will prevent extension.

Excessive movement in the frontal plane may occur as a result of muscle weakness; for example, the hip (femur) may adduct through the loading response phase if the gluteal muscles are weak and unable to control that motion. This will vary between individuals and may be affected by things such as the width of the pelvis.

The knee also moves in the sagittal plane, and this can be observed through simple gait analysis. Again,

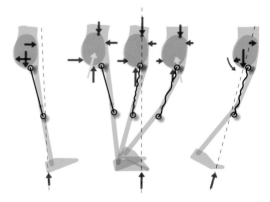

▲ HIP AND KNEE ANGLES

The angles of hip and knee during a normal walking gait.

the degree of this flexion and extension will be dependent on other factors, including step length, speed, range of motion of the knee, and accessory motion at the knee. At initial contact, the knee will be almost in full extension unless restricted by something such as pain. The degree of knee flexion will depend on a range of factors, including the range of motion in other joints such as the hip and ankle. If these joints are restricted,

there may be compensations such as increased motion, which may then cause problems at the knee. It could therefore be advocated that those suffering with knee pain due to running should consider gait analysis from an appropriately qualified practitioner.

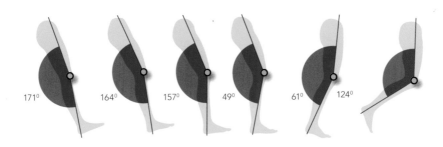

171°   164°   157°   49°   61°   124°

▲ KNEE FLEXION DURING WALKING

The angle of the knee during a walking gait.

Uphill and downhill running

RUNNERS WILL OFTEN ADD TO THEIR
TRAINING BY INCLUDING HILL
RUNNING, WHICH WILL WORK THEIR
BODIES IN DIFFERENT WAYS AND PUT
DIFFERENT STRESSES ON THE JOINTS
AND MUSCLES. The images of uphill
and downhill running provide a simple
view of how the angle of the upper
body changes when running uphill or
downhill to counter the forces being
placed on them. When running uphill,
the trunk tends to be in a more flexed
position, and the hips must flex more

with a higher knee lift in order to clear
the ground and move upward.

When running downhill, it can be
argued that there are greater stresses
on the body, particularly the muscles.
Since momentum will carry an
individual downhill, muscles have to
work eccentrically to control speed
and prevent body weight from moving
forward too much. This eccentric load
on the muscles will help to strengthen
them, but stress on the muscle fibers
as a result of downhill running is one

◀ UPHILL
The knee tends to be more flexed
during a run uphill.

reason why runners suffer from muscle soreness after this type of exercise. When running downhill, the trunk is more upright to counter the momentum of the body moving forward, and this shifts the center of gravity, resulting in different forces acting on the body. Both uphill and downhill running can be beneficial for runners since the loading is different and can improve the strength and endurance of a runner. However, these types of sessions should be introduced gradually to allow the body to become accustomed to the different forces being placed on it and for the muscles to be trained to cope with this type of activity.

◀ DOWNHILL

Running downhill exerts an eccentric strain on the muscles of the lower limb.

Barefoot running

In recent years there has been a trend toward barefoot running, with training shoe companies designing shoes to replicate this. Proponents of barefoot running advocate that it reduces the risk of injury, can be more efficient, and improves running performance. One principle that might explain this is the relationship between ground reaction force and absorption of force by different joints.

Running barefoot promotes a forefoot strike, as seen in the illustration, far right, and it occurs quite naturally when barefoot.

This style of running results in a more upright stance and promotes greater flexion in the knee and hip to absorb the forces, which would otherwise be placed on the knee when it is in an extended position. It is argued that this provides a mechanical advantage because the muscles of the lower limb are able to withstand the forces, rather than the joints themselves having to absorb them, which can lead to damage and injury.

There is evidence that some running-related injuries might be due to overpronation. Symptoms

▶ **HIGH-IMPACT STYLE**

Proponents of the barefoot running style argue that there are reduced forces on the body, as seen in the image on the right. They argue that heel-strike gait causes greater impact on an extended lower limb.

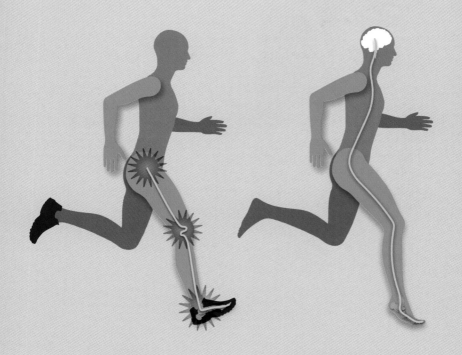

HEEL STRIKE

FOREFOOT STRIKE

Barefoot running *(cont.)*

of overpronation include medial tibial stress syndrome (shin splints) and Achilles tendinopathy. Running with a forefoot strike could potentially eliminate these types of injuries. The greater absorption of forces by the lower limb muscles would also reduce the stress on the joints of the lower limb and on the lower back.

Walking barefoot may also have benefits, despite the fact that there may still be a heel strike.

Walking barefoot should promote a shorter stride length, allowing a softer landing on the heel. As loading response occurs, the foot can pronate more easily because a shoe does not restrict it, and therefore, the foot is allowed to spread out and withstand the ground forces. Walking barefoot also strengthens the intrinsic foot muscles and the other structures of the foot, which should be of benefit when requiring the foot to

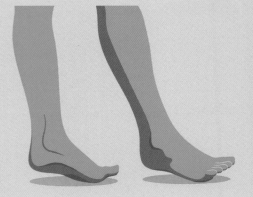

▶ **BAREFOOT WALKING**

Walking barefoot, it has been argued, is beneficial because the greater surface area can absorb more force. Some argue that physiologically this is what we are programmed to do.

become more rigid for toe off. There are potential risks with barefoot running, and some injuries may become more prevalent. An obvious issue with running or walking barefoot is linked to the environment, and doing this on hard surfaces such as concrete can cause blisters and painful skin reactions. Further injury risks include stress fractures to the metatarsal bones as they contact the ground with a forefoot strike. The muscles of the lower limb will also be worked differently and may be subject to damage as they struggle to adjust to the increased demands placed upon them. For example, many runners who try a forefoot strike style of running will often complain of

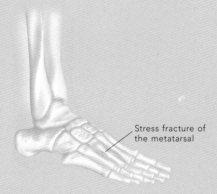

Stress fracture of
the metatarsal

▶ **STRESS FRACTURE**
Stress fractures to the metatarsals are injuries often associated with barefoot running.

Barefoot running *(cont.)*

soreness in their calf muscles, which are loaded more than in the heel strike style of running.

Therefore, there appear to be advantages and disadvantages to barefoot running, and anyone thinking about changing to this style of running should do so gradually and consider the environment where they intend to run.

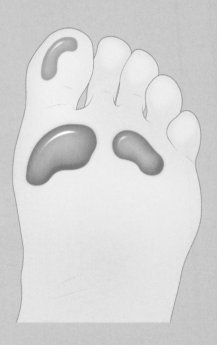

▶ **BLISTERS**

Painful skin reactions, such as blisters, are a possible side effect of walking barefoot.

BAREFOOT STRIKE

Shorter strides allow you to lean softly on your heel with your knees slightly bent—the way you might walk on a beach.

A natural step through the outside edge of the foot, before the ball lands—and spreads slightly—on the ground.

Your toes are designed to give you a powerful push forward—sending you striding smoothly into the next step.

Normal ranges and values

Nervous and sensory systems

Conduction velocity of large myelinated fiber: 80 to 120 m/s
Conduction velocity of small unmyelinated fiber: 0.5 to 2 m/s
Spinal segments supplying the diaphragm: C3 to C5
Spinal segments supplying the upper limb: C5 to T1
Spinal segments supplying the sympathetic outflow: T1 to L1
Spinal segments supplying the lower limb: L2 to S3
Spinal segments supplying the bowel and bladder: S2 to S4
Parasympathetic outflow: Cranial nerves 3, 7, 9, and 10; sacral segments 2 to 4
Near point of eye (closest distance that can be focused on): 100 mm at age 25
Frequency range of normal human hearing: 20 Hz to 20,000 Hz
Decibel levels (normal conversation): 60 to 70 dB

Cardiovascular system

Heart rate
  At rest: 60 to 70 beats per minute
  Maximal exercise: 200 bpi
Stroke volume
  At rest: 70 ml
  During exercise: 200 ml
Blood pressure at rest
  Systolic: 120 mm Hg
  Diastolic: 80 mm Hg
Blood pressure during exercise
  Systolic: 180 mm Hg
  Diastolic: 85 mm Hg
Blood flow
  Rest: 20% to skeletal muscle
  Exercise: 80% to skeletal muscle
Total blood volume: 5 liters

Respiratory system

Breathing frequency (respiration rate)
  Rest: 12 breaths per min.
  Max. exercise: 30 breaths per min.
Tidal volume: 500 ml
Anatomical dead space: 150 ml
pO_2 air: 160 mm Hg
pCO_2 air: 0.3 mm Hg
pO_2 alveoli: 105 mm Hg
pCO_2 alveoli: 36 mm Hg
Oxygen uptake:
  At rest: 3 ml/kg/min
  Max exercise: 70 ml/kg/min

Maximal oxygen uptake: Men

Rating	Age (years)					
	18–25	26–35	36–45	46–55	56–65	65+
excellent	>60	>56	>51	>45	>41	>37
good	52–60	49–56	43–51	39–45	36–41	33–37
above average	47–51	43–48	39–42	36–38	32–35	29–32
average	42–46	40–42	35–38	32–35	30–31	26–28
below average	37–41	35–39	31–34	29–31	26–29	22–25
poor	30–36	30–34	26–30	25–28	22–25	20–21
very poor	>30	>30	>26	>25	>22	>20

Maximal oxygen uptake: Women

Rating	Age (years)					
	18–25	26–35	36–45	46–55	56–65	65+
excellent	>56	>52	>45	>40	>37	>32
good	47–56	45–52	38–45	34–40	32–37	28–32
above average	42–46	39–44	34–37	31–33	28–31	25–27
average	38–41	35–38	31–33	28–30	25–27	22–24
below average	33–37	31–34	27–30	25–27	22–24	19–21
poor	28–32	26–30	22–26	20–24	18–21	17–18
very poor	>28	>26	>22	>20	>18	>17

Gastrointestinal system

pH of stomach juices: 1.5 to 3.5
Transit time of gastrointestinal tract: 18 hours to 3 days

Urinary system

Glomerular filtration rate: 125 ml/min.
Tubular reabsorption rate: 124 ml/min.
Urine output: 1 ml/min or approximately 1.5 L/day

Composition of urine

Osmotic concentration: 850 to 1340 mOsm/L
Specific gravity: 1.003 to 1.030
pH: 4.5 to 8.0, with a mean of 6.0
Bacterial content: nil, urine should be sterile
Red blood cells: 100/mL
White blood cells: 500/mL
Sodium: 330 mg/dL
Potassium: 166 mg/dL
Chloride: 530 mg/dL
Calcium: 17 mg/dL
Urea: 1.8 g/dL
Creatinine: 150 mg/dL
Ammonia: 60 mg/dL
Uric acid: 40 mg/dL
Urobilin (yellow pigment): 125 µg/dL

Blood chemistry

pO_2 of systemic arterial blood: 75 to 100 mm Hg

pCO_2 of systemic arterial blood: 35 to 45 mm Hg

Sodium: 138 mM

Potassium: 4.4 mM

Chloride: 106 mM

Bicarbonate: 27 mM

Blood pH

   Rest: pH 7.4

   Maximal exercise: pH 7.1

Urea: 10 to 20 mg/dL

Creatinine: 1 to 1.5 mg/dL

Ammonia: <0.1 mg/dL

Albumin: 3.6 to 4.7 g/dL

Glucose (whole blood, fasting): 3.3 to 5.6 mM (60 to 100 mg/dL)

Hemoglobin

   Male: 13.8 to 18.0 g/dL

   Female: 12.1 to 15.1 g/dL

Hematocrit (proportion of total blood volume)

   Male: 38 to 54%

   Female: 35 to 48%

Cellular composition

Red blood cell mean volume: 80 to 100 fL (femtoliters 10–15)
Red blood cell mean Hb concentration: 310 to 360 g/L
Total leukocyte population: 4.0 to 11.0 x 109/L
Neutrophil population: 2.0 to 7.5 x 109/L
Lymphocyte population: 1.0 to 4.0 x 109/L
Monocyte population: 0 to 1.0 x 109/L
Eosinophil population: 0 to 0.5 x 109/L
Basophil population: 0 to 0.3 x 109/L
Platelet population: 150 to 450 x 109/L

Strength loss

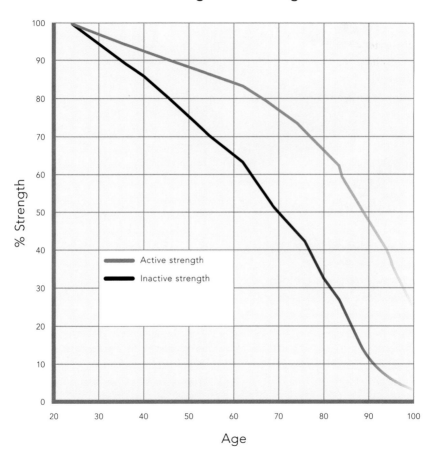

Strength loss with age

% Strength (y-axis): 0, 10, 20, 30, 40, 50, 60, 70, 80, 90, 100

Age (x-axis): 20, 30, 40, 50, 60, 70, 80, 90, 100

Legend:
- Active strength
- Inactive strength

Index

Acknowledgments

With thanks to the contributing authors for this book: Elaine Mullally MSc BSc (Hons) FHEA GSR, Senior Lecturer in Sport Rehabilitation at St. Mary's University, UK (chapters 6, 8, 12, 13, and 15); Oliver Blenkinsop BSc GSR MSc AT, Lecturer in Sport Rehabilitation at St. Mary's University, UK (chapters 5, 8, 12, and 13); Nic Perrem, Lecturer in Sport Rehabilitation at St. Mary's University, UK (chapters 7, 8, and 12–14); Dr. Michael Baker PhD ESSAM AES AEP, Senior Lecturer, School of Exercise Science at Australian Catholic University (chapters 1–4, 9, and 10).